U0079382

商業
實用英文
E-mail

業務篇
附文字光碟

國家圖書館出版品預行編目資料

商業實用英文E-mail・業務篇／張瑜凌編著

-- 三版 -- 新北市：雅典文化，民108.03

面； 公分 --（全民學英文；52）

ISBN 978-986-96973-6-1(平裝附光碟片)

1. 商業書信 2. 商業英文 3. 商業應用文 4. 電子郵件

493.6　　　　　　　　　108000100

全民學英文系列 52

商業實用英文E-mail・業務篇

編著／張瑜凌
責任編輯／張瑜凌
美術編輯／王國卿
封面設計／林鈺恆

法律顧問：方圓法律事務所／涂成樞律師

總經銷：永續圖書有限公司　　CVS代理／美璟文化有限公司
永續圖書線上購物網　　　　 TEL：(02) 2723-9968
www.foreverbooks.com.tw　　FAX：(02) 2723-9668

出版日／2019年03月

雅典文化

出版社　22103　新北市汐止區大同路三段194號9樓之1
TEL　(02) 8647-3663
FAX　(02) 8647-3660

序言 不要忽視「E-mail 商業實用英文」的影響力

在電子科技日新月異的廿一世紀，人與人之間的溝通早已從飛鴿傳書的書信往返時代跨越至利用 e-mail 作為主要的溝通工具；再加上地球村世代的來臨，「英文」成為共通的國際語言已是不爭的事實了。因此，當你面對必須和國外客戶聯絡的機會時，別讓看似困難其實簡單的英文壞了您的大事。

想要擁有競爭力，首先你必須同時具備兩種能力：利用 e-mail 發信及商用英文寫作。

要寫出一封合宜得體又發揮商業行銷功效的「英文商用書信」(Business English)，您只需要具備基本的寫作技巧，本書特別針對 e-mail 初級入門學習者及商用書信學習者兩大族群，提供了五大學習技巧：

一、【撰寫 e-mail 英文商用書信的訣竅】

二、【常用商業例句】

三、【常用商業名詞】

四、【商業書信範例】

五、【商業常用詞彙】

針對不同的商業溝通情境，逐一介紹、演

練，並提供多種簡易的術語及基本語法、關鍵片語，讓您在最短的時間具備英文書信寫作的技巧，輕輕鬆鬆學會如何寫出一封正確、得體的商用英文書信。

本書從基礎的 e-mail 格式介紹、信函的書寫格式及各類型的應用實例，深入淺出，一步一步引領您克服英文寫作的恐懼，英文雖然不是您的母語，但是透過正確的學習管道，您一樣可以應用自如，若加上 e-mail 電子媒介無遠弗屆的傳播力量，更能發揮英文商用文書的影響力。

「具備實力才有競爭力」，面對多變的、科技的、地球村的時代，您準備好面對挑戰了嗎？

「商業實用英文 E-mail」學習手冊

網際網路的崛起，直接地改變了生活、職場上的溝通型態，過去透過電話、書信、傳真或出差等的溝通模式，如今又多了另一個更為直接、有效率的溝通方法：電子郵件 e-mail，也促使「地球村」概念的日趨成形。

如何能夠寫出一封有效率的電子郵件呢？您必須要瞭解電子郵件的形成要素。

本書透過循序漸進的說明介紹，讓您能夠快速地進入電子郵件的溝通世界。

一、撰寫 e-mail 英文商用書信的訣竅

Tip

1 常見的英文日期表達方式：

在英文書信中，常見的日期表達方式，有以下兩種方式：

(1) "mm/dd/yy"，表示「月份/日期/年份」，如"05/16/2016"

(2) "dd/mm/yy"，表示「日期/月份/年份」，如"16/05/2016"

本書從基礎的 e-mail 格式介紹、信函的書寫格式及各類型的應用實例，深入淺出，一步一步引領您順利完成一封商業 e-mail。只要您能按部就班地學習，英文寫作其實沒有想像中那麼難以掌握。

二、常用商業例句

一、簡單開場白

例 We are desirous of extending our connections in your country.

我們擬拓展本公司在貴國的業務。

本單元提供上百句實用的商業書信例句，針

對不同的情境供您選擇使用，包含：自我介紹、會面、詢價、樣品、訂單、付款等單元，讓您能夠從容地應付各種突發狀況。

三、商業常用名詞

1. make an inquiry	詢價
2. credit enquiry	信用查詢
3. credit status	信用狀況
4. creditability	信用
5. financial standing	財務狀況
6. mode of doing business	交易方式
7. in detail	細節
8. products[productions]、goods	產品
9. price list	價格明細
10. import[export] price	進[出]口價
11. sample	樣品
12. terms of payment	付款方式
13. time[date] of delivery	交貨時間[日期]

　　想要成為一名專業的商業行銷菁英，除了擁有良好的溝通能力之外，您還必須具備足夠的專業知識，本單元提供給您相關的商用常用名詞，讓您能夠正確無誤地與外國客戶溝通，以免用錯字句造成貽笑大方的窘境。

四、商業書信範例

原文範例

Gentlemen,

We thank you for your **cooperation** for our business for the past five years.

Now we are desirous of enlarging our trade in staple **commodities**, but have had no good connections in Canada.

Therefore we shall be **obliged** if you could kindly **introduce** us to some of reliable **importers** in Canada who are **interested** in these lines of goods.

We await your **immediate** reply.

　　除了有「原文範例」、「翻譯範例」及「關鍵單字」之外，還包含範例的詳細說明及關鍵片語，讓您能夠熟讀之後，發揮最大的 e-mail 溝通效率，使您能夠在短時間內就寫好一封 e-mail。

此外，「原文範例」就提供您一個解決的方式，您可以依照原文撰寫，再依不同的需要替換其中的「關鍵單字」即可。

五、商業常用詞彙

編列常用的商務貿易相關詞彙，分門別類整理規劃，讓您方便查詢使用。

03 常用商業名詞

04 商業書信範例

05 商業常用詞彙

Chapter

1

撰寫 e-mail
英文商用書信的訣竅

在二十一世紀，因為在網際網路的盛行，使人們越來越依賴電子郵件（e-mail）的溝通模式，似乎除了電話之外，e-mail 已成為第二順位的溝通選擇，因為電子郵件除了基本的「溝通」特點之外，還兼具提醒、通知、條列式說明的特色。

在職場上，想要寫好一篇得體又專業的商用 e-mail 書信，您必須要有幾項基本能力：

一、通曉商業術語及慣用語

二、優秀的英文文法、句型的基礎

三、得體的書面禮儀

本書就是教您如何善用電子郵件，讓 e-mail 成為您無往不利的商務溝通工具。

一般而言，一封 e-mail 包含以下幾個主題：

一、 Heading 信頭

二、 Salutation 稱謂

三、 Body of letter 正文

四、 Terms of respect 敬辭

五、 Signature 簽署

六、 Postscript 附註說明

在上述主題中，各自有不同的表現方式，您必須先瞭解其中的內容定位及書寫步驟。

您可以在下一頁看見一封商業書信的 e-mail 範例，幫助您更加瞭解商務用英文書信的基礎結構。

※商業 e-mail 結構範例

信頭

Date Sent: Tue, 5 May 2016
From: Vivian <vivian@yahoo.com>
Subject: Our new order
To: Chris <chris@yahoo.com>
Cc: Maria <maria@yahoo.com>

稱謂

Dear Chris,

正文

We have not received your order since last October. We need to find out what the trouble has been.

It is our policy to render the best service to customers. You have always been considered as one of our regular customers, for you have given us remarkable patronage. We hope to have the pleasure of serving you again soon. Your kind reply will be much appreciative.

敬辭

Sincerely yours,

簽署

Vivian Wang

附註

P.S. We need you and we do not want to lose your business.

一、Heading「信頭」

想要利用 e-mail 作為商用書信的溝通管道時，首先你必須要瞭解每一種 e-mail 工具所代表的意義，在撰寫商用書信時，才能更事半功倍地發揮其功效。

在一般的 e-mail 發信軟體中，都會有一制定的 "Heading"(信頭)格式，雖然各家郵件軟體的格式不盡相同，但是所提供的功能是大同小異的，您可以依照個人的使用習慣或公司的安排，使用不同的郵件軟體。

以下這些信頭的內容代表的意義皆不同，您不得不了解其重要性：

(一) Date Sent「寄件日期」
(二) From「寄件者」
(三) Subject「主旨」
(四) To「收件者」
(五) CC「副本」
(六) BCC「密件副本」
(七) Attached「附件」

(一) Date Sent「寄件日期」

"Date Sent"代表你所發信件的日期。已知目前市面上常用的電子郵件軟體(例如 outlook、outlook express、yahoo、hotmail、pegasus)幾乎都會在撰寫郵件時,自動加上發信當時的日期,有的電子郵件軟體甚至還會增加發信的時間。

雖然你不必親自註明日期,但是你還是要瞭解在書寫「商用書信」時,「英文日期」的表達方式:

1. 日期的表達方式:

一般美式英文日期的書寫方式為:"Tue, May 5, 2016"表示是在「二〇一六年五月五日星期二」所發的信函。

Tip

❶ 常見的英文日期表達方式:

在英文書信中,常見的日期表達方式,有以下兩種方式:

(1) "mm/dd/yy",表示「月份/日期/年份」,如"05/16/2016"

(2) "dd/mm/yy",表示「日期/月份/年份」,如"16/05/2016"

Tip
② month「月份」的表達方式容易混淆：

其中，因為語言使用的慣例不同，又以「月份」和「日期」最容易為東方人所誤解。

因此建議您在書寫商用書信時，若要特別提及日期時，「月份」最好不要用阿拉伯數字表示，而改以月份的英文縮寫表達。(詳見第 24 頁「日期的相關單字」的説明)

Tip
③ 有關於「日期順序」的表達方式：

在正式英文商業書信中，日期的使用順序分別為「星期」、「月份」、「日期」、「年份」，例如「二〇一六年十二月廿日星期四」有以下三種書寫順序：

(1) Thu, December 20, 2016
(2) Thu, December 20th, 2016
(3) December 20, 2016（不含星期）

2. 日期的相關單字：

(1) weekday & weekend 工作日和週末：

因為東西方對於語言使用文化上的差異，所

以常常造成的困擾之一，就是對於"next"的使用時機的差異性。舉例來説：

若今天是一月一日星期日，那麼在"next Monday"的解讀上，便會產生以下的差異：

東方人的"next Monday"，解讀為「一月二日的星期一」。

西方人的"next Monday"，解讀為「一月九日的星期一」。

所以建議您，當要表示星期的日期時，還是應該夠另外再增加日期的説明。

英文	中文	縮寫
Monday	星期一	Mon.
Tuesday	星期二	Tue.
Wednesday	星期三	Wed.
Thursday	星期四	Thu.
Friday	星期五	Fri.
Saturday	星期六	Sat.
Sunday	星期日	Sun.
weekday	工作日 （星期一至星期五）	
weekend	週末 （星期六及星期日）	

(2) month 月份:

雖然月份也可以用阿拉伯數字表示,但是為了避免在溝通過程中,造成彼此任何的混淆或誤解,建議還是用正規的英文名稱或縮寫來表達月份。

英文	中文	縮寫
January	一月	Jan.
February	二月	Feb.
March	三月	Mar.
April	四月	Apr.
May	五月	May
June	六月	Jun.
July	七月	Jul.
August	八月	Aug.
September	九月	Sep.
October	十月	Oct.
November	十一月	Nov.
December	十二月	Dec.

(3) date 日期：

可以直接用數字表示，如「五月二日」就是用" 2, May" 或是"the 2 of May"表示。但在正式英文中，月份的日期是用序號表示，例如上述句子，為了慎重起見，建議您用正式的日期"the second of May" 表達。

英文	中文	縮寫
first	第一	1st
second	第二	2nd
third	第三	3rd
fourth	第四	4th
fifth	第五	5th
sixth	第六	6th
seventh	第七	7th
eighth	第八	8th
ninth	第九	9th
tenth	第十	10th
eleventh	第十一	11th
twelfth	第十二	12th
thirteenth	第十三	13th
fourteenth	第十四	14th
fifteenth	第十五	15th
sixteenth	第十六	16th
seventeenth	第十七	17th
eighteenth	第十八	18th
nineteenth	第十九	19th
twentieth	第廿	20th

英文	中文	縮寫
twenty-first	第廿一	21st
twenty-second	第廿二	22nd
twenty-third	第廿三	23rd
twenty-fourth	第廿四	24th
twenty-fifth	第廿五	25th
twenty-sixth	第廿六	26th
twenty-seventh	第廿七	27th
twenty-eighth	第廿八	28th
twenty-ninth	第廿九	29th
thirtieth	第卅	30th
thirty-first	第卅一	31st

(4) year 年份：

　　直接用數字表示，如"2016" 表示「二〇一六年」，通常年份是放在整句句子的最後，比較沒有爭議。

T i p

④ 「2016 年」的英文怎麼說？

1995 的英文是"nineteen ninety-five"，字面解釋是「十九」加上「九十五」，那麼「2016 年」怎麼說呢？可不是"twenty sixteen"，正確的說法應是："two thousand and sixteen"，也就是「二千」加上「十六」的說法，可不要說錯喔！

　　以下再總整理出日期的說明例句，您可以依實際需要改寫：

☑ We will place an order before September of 30, 2016.

☑ We will place an order before September 30th, 2016.

☑ We will place an order before Sep. 30, 2016.

☑ We will place an order before the 30th of September.

☑ We will place an order before Monday, Sep. 30, 2016.

(二) From「寄件者」

"from"是「從~（何處）而來」的意思，而非絕對代表「發信人」（sender），這是欄位的註明方式。

和「日期」一樣，在撰寫郵件時，發信軟體通常會自動加上你所註冊的帳號或顯示的名字。

若你是代表公司的名義撰寫商用書信，建議應以讓對方在第一時間就瞭解你(或企業)的身分為「寄件者」的顯示設定，以免被收件者誤當成「垃圾郵件」(spam) 而刪除(delete)，反而讓商務信件遭受石沉大海的命運。

因為"from" 的內容可以依每個人各自的設定而有不同的顯示名字，您可以需求，選擇以下的顯示方式：

Tip

❶ 表示「個人」的寄件者

"From:"Vivan" vivan@yahoo.com"
表示「個人名字」＋「郵件地址」。

T i P
② 表示「身分全名」的寄件者

"From: Chris Nietopski"

　「來自克里斯・尼達布思金」，表示「名字+姓氏」。

T i P
③ 表示「公司代表」的寄件者

"From: Chris of BCQ"

　「來自 BCQ 公司的克里斯」，表示「個人名字」+「公司名字」。

T i P
④ 表示「所屬企業部門」的寄件者

"From: BCQ Inc. Customer Service"

　「來自 BCQ 企業客戶服務部門」，表示「公司部門身分」。

T i P
⑤ 表示「公司身分」的寄件者

"From: the BCQ Product Upgrade Team"

　「來自 BCQ 公司產品升級小組」，表示「公司組織的身分」。

E-MAIL

(三) Subject「主旨」

"subject"表示「主旨」，代表這封商用書信的「主題」。

英文中有一句話叫做"Time is money"（時間就是金錢），因此在分秒必爭的商業行為中，「時間」更是非常珍貴的。

而一封「有效率」（efficiency）的商用書信，其「主旨」（subject）欄位所傳達的內容，更是能夠幫助「收信人」（receiver）對信件的篩選產生加分的作用，自然能夠讓你的e-mail商用書信能夠在一堆「垃圾郵件」（spam）中脫穎而出。

以下是一些在商用書信中常見的「主旨」說明，提供您簡化書信內容的解釋：

Tip ①

1. Quotation	報價單、估價單
2. Offer	出價、報價
3. Bargain	議價
4. Budget	預算
5. Sample	樣品
6. Catalogue	型錄

7. Order/Purchase order	訂單
8. Contract	合約
9. Shipment	裝船(運)
10. Damage	損壞
11. Shortage	數量短缺
12. Complaint	抱怨
13. Reminder	提醒
14. Agency	代理權
15. Payment	付款
16. Inquiry	詢問
17. Request	需求
18. Information	資訊
19. Follow-up	後續
20. Detail	細節

　　此外，若是覺得簡短的主旨無法充份傳達您想要表示的意思時，您也可以利用較長的主旨說明，以提醒收信者注意您的商用書信，但是以不超過五個單字的簡易句子為原則，以免過於冗長，而讓對方失去看信的耐性：

Tip
②

1. Making an appointment	安排會議
2. Looking for Customers	尋找客戶
3. Extending Business	拓展業務
4. Trade Proposal	商務計畫
5. Agreement on Conditions of Business	
	交易條件協議
6. Trade Inquiry	商務詢價
7. Selling Offer	賣方報價
8. Buying Offer	買方報價
9. Placing an Order	下訂單
10. Follow-up	進度查詢
11. Sales Contract	買賣合約
12. Claims and Adjustments	索賠與調處
13. Establishment of Agency Ship	
	建立代理關係

　　您也可以用 "Sub" 或 "Re" 表示主旨的說明文開頭，前者是 "Subject" 的縮寫，後者是 "Reply" 的縮寫，適用在「回覆」對方前一封信件的意思。

Tip ③ 主旨＋説明

　　商用書信的主旨內容以「主旨」＋「日期」或是「主旨」＋「説明」的方式表示，例如：

1. Sub: Your letter of Jan. 30, 2016
   主旨：您二〇一六年一月卅日的來信
2. Sub: Your Order No. 203
   主旨：您編號 203 的訂單
3. Sub: Inquiry from BCQ
   主旨：來自 BCQ 的詢問

Tip ④ 回覆＋説明

1. Re: Shortage of your goods
   回覆：您短缺的商品
2. Re: Sorry for being late
   回覆：為遲到而抱歉
3. Re: Update of product list
   回覆：更新產品清單

(四) Receiver「收件人」

　　"receiver"代表這封商用書信的寄發(send)對象。商用書信中,你必須有一特定的寄件對象,在你選擇對方的郵件地址之後,對方的名字(或郵件地址)就會出現在此欄位中。

　　一般而言,每一種寄信軟體都有「通訊錄」(address book),在設定「收件人姓名」的內容時,就可以在其名字之後,直接註明對方的公司、職銜,而郵件地址(e-mail address)就鍵入至相關的欄位中,當然也可以僅用其名字顯示即可,但是前提必須是此郵件軟體有「顯示名稱」的選項,例如 Hotmail 軟體。

　　以下為幾種「給收件人」的表現方式:

①

To:"Mr. White" (給「懷特先生」),直接表示收信者姓名。

②

To:"Chris" chris@yahoo.com (給「帳號 chris@yahoo.com 的克里斯」),表示「個人名字」+「郵件地址」

Tip
3

To:"Chris" of BCQ (給「BCQ 公司的克里斯」)，表示「個人名字」+「公司名字」

Tip
4

To:"Chris Jones" (給「克里斯・瓊斯」)，表示「名字+姓氏」。

Tip
5

To:"Manager of Marketing Department" (給「行銷部門經理」)，表示「職銜」。

(五) **CC**「副本」

當需要將此信寄給第三者時，就可以在"CC"的欄位中鍵入第三者的郵件地址，而主要的收信人也會知道你同時有給第三者這一份副本。

"CC"的全文是"Carbon Copy"。"Carbon"原意為「碳」，因為以前的複寫紙都含有碳，所以「複寫的副本」的英文就叫做 "Carbon Copy"。

❶ 「副本」怎麼使用？

　　若是你需要對方發信件時，能夠同時提供一份副本怎麼説？很簡單，將 "CC" 的縮寫當成動詞即可，雖不是正式的文法，但也可以充分傳達發送副本的意思：

✷ If you have any idea, please CC me a copy.
　如果你有任何的想法，請發副本給我。

✷ Please CC me at lisa@yahoo.com
　請發副本到 lisa@yahoo.com 信箱給我。

✷ FYI, if you have any comments, please CC me.
　給您參考，假使您有任何建議，請發副本給我。

✷ Please CC me when you make contact with anyone.
　當你和任何人聯絡時，請發副本通知我。

(六) BCC「密件副本」

　　若是你需要將此信寄給第三者，但又不想讓主要收信人知道有第三者也收到這封信，您就可以在 "BCC" 的欄位中鍵入第三者的郵件地址。

"BCC" 的全文是"Blind Carbon Copy"。

"Blind"是表示「視而不見的」，藉此表示收件者看不見副本收件人，也就成為「密件副本」的意思。

(七) Attached file「附件」

除了選取「附加檔案」功能加入檔案之外，你也可以在內文當中說明您有隨信附加檔案給對方，以提醒對方開啟附加檔案。

T i p

❶ 「請參考附件檔案」的說明

若是擔心對方忽略了您所附加的檔案，那麼在信件的最後，你就可以提醒對方：

Please see attached file for more detailed information.

更多詳情，請詳見附加檔案。

二、 Salutation「稱謂」

"Salutation"「稱謂」是屬於e-mail信件中，在書寫正文前的第一部份，也就是中文書信「敬啟者」的意思。

"Salutation"「稱謂」代表對收件人的尊稱，就如同人見面要打招呼一樣，您不可不注意在此時應注重的禮儀，特別是商用書信中，你的所有稱呼用詞都攸關著是否能成功地利用「e-mail 商用書信」作為溝通工具。

由於英文是東西向橫寫文字，「稱謂」的書寫位置便在信件本文的第一行，必須靠齊內文的最左側。

不管是正式的或非正式的商用書信，你對對方的稱呼，都應以尊敬對方、與對方保持良好關係為前提。

以下是一些商業書信中常用的稱謂方式：

(一) **Dear Sir**「敬啟者」：

用在「只知其人不知其名」的狀況下，例如你想要向某公司的「客戶服務部門」投訴時，因為你不會知道是哪一個特定人員閱讀此電子郵件，就可以用此稱謂尊稱收信者。

(二) **My dear Sir**「敬啟者」：

也是用在「只知其人不知其名」的狀況下，使用情況同上述，但是多了"My dear"的稱呼，便多了一份尊敬、謙虛的態度。

(三) **Dear Sirs**「敬啟者」：

用在只知為一群人中的「某一人」，使用情況也類似上述情形，但是你確知那可能會是「很多人」都會收到訊息的情況下時使用。

Tip

❶ Sir 的性別？

一般而言，Sir 多半適用在男性對象，但是若是不知收信者的性別時，有時也會統一使用 Sir 的尊稱，此時便不必在意對方的性別，也可以適用，但必須注意，使用前提是對方是你所不認識者才能使用。

(四) **Dear Madam**「親愛的小姐」：

"Madam" 表示「夫人」、「太太」、「小姐」的尊稱，適用對象不分已婚或未婚，也不知該名女士的稱呼，但是表示你已知收信者為「女性」，但不知道對方的身分或姓名，也無從得知是哪一位特定女性會閱讀郵件，是正式書信中，對女性的最常用的尊稱，後面不加任何的姓氏或名字。

(五) **Dear Sir [Madam]**「敬啟者」：

在書寫時，男女的性別都註明時，就有點類似中文所説的「先生或小姐」的意思，也是適用在不知收信者性別的情況下使用，因此不論收信者是哪一個性別，都不會顯得突兀、不禮貌。

(六) **Dear Mr. [Mrs.] Jones**「親愛的瓊斯先生 [瓊斯太太]」：

"Mr." 是"Mister" 的簡稱，而"Mrs." 則為"Mistress"的簡稱。

"Mr./Mrs. +姓氏（last name）" 用在僅知道對方的姓氏時。適用於所有正式場合或非正式場合對男性的尊稱。

但是對女性使用"Mrs." 則是確認對方為「已婚身分」，並是冠上「丈夫姓氏」時使用。

若是您不知對方女性是否已婚或是否冠上先生的姓氏，建議您還是用"Dear Madam" 比較不會造成雙方的誤解。

Tip
② Mr./Mrs./Ms.的用法

(O)Mr.+姓氏
　例如：Mr. Jones（瓊斯先生）

(O) Mr.+名字+姓氏

例如：Mr. John Jones（約翰·瓊斯 先生）

(X) Mr.+名字

例如：Mr. John（約翰先生）

(七) Dear Mr. and Mrs. Jones「親愛的瓊斯先生及瓊斯太太」：

代表收信者是一對已婚的夫妻，並且你的信件內容與其夫妻都有關係時適用，像是邀請夫妻一同參加聚會、共同通知夫妻雙方某事時，就可以使用" Dear Mr. and Mrs. Jones"。

T i p

3 Mr. and Mrs. 夫婦的用法

(O) Mr. and Mrs.+姓氏

例如：Mr. and Mrs. Jones（瓊斯夫婦）

(O) Mr. and Mrs.+先生名字+姓氏

例如：Mr. and Mrs. John Jones（約翰·瓊斯夫婦）

(X) Mr. and Mrs.+太太名字+姓氏

例如：Mr. and Mrs. Mary Jones（瑪莉·瓊斯夫婦）

(八) **Dear Miss Jones**「親愛的瓊斯小姐」：

若是對方為未婚的女性，並知道其姓氏時，就可以使用"Miss+姓氏" 稱呼對方，表示對未婚女士的稱呼。此外，有時明知對方為已婚身分，也可以用"Miss+（女方）姓氏"的方式稱呼對方，但似乎會顯得刻意逢迎的態度。

④ Mr. and Mrs.的動詞

(O) Mr. and Mrs.+姓氏+複數動詞

例如：Mr. and Mrs. Jones were married on September 25, 1965.

（瓊斯夫婦是在一九六五年九月廿五日結婚。）

(九) **Dear Ms. Jones**「親愛的瓊斯女士」：

當你已知對方身分，但表示無法確定對方是否已婚或未婚的女性身分，或對方為不願提及婚姻狀況的女性時，就可以用"Ms.+姓氏"來稱呼對方，簡而言之，就是已婚、未婚女士皆適用。

(十) Dear Doctor Jones「親愛的瓊斯醫師」：

當你已知對方的職銜時，就可以用"職銜+姓氏"稱呼對方。無關乎男女性別或是否已婚的身分，"職銜+姓氏"的稱謂方式都適用在商用書信中，而此時職銜第一個字母通常都使用大寫。

(十一) Dear Customer「親愛的客戶」：

當只知對方身分但沒有特定哪一位對象時，就可以使用"Dear+身分"的稱謂，通常是對對方表示無上的敬意時使用，像是給顧客（Customer）、股東（Stockholder）、會員（Member）、出席者（Attendant）、員工（Employee）的信件都可以採用"Dear+身分"的方式，此種信函也有類似「通知函」的功能；同上述職銜，第一個字母也是大寫。

(十二) Gentlemen「敬啟者」：

使用情況和"Dear Sirs"類似，適用在你的寫信目標為只知是一群人中的「某一人」時，在商業書信中的使用非常普遍。

"Gentlemen"是「紳士」(Gentleman)的複數名詞。

(十三) To whom it may concern「貴寶號鈞鑒」：

　　當你寫信給某一公司或單位時，在不知道收件者是某位特定人、特定部門時，就可以使用"To whom it may concern"，字面意思是表示「給予此事相關者」，即為「貴寶號鈞鑒」或「敬啟者」的意思。

(十四) Dear Chris「親愛的克里斯」：

　　直接稱呼對方"Dear+名字"是使用在非常熟悉對方的情況下，其熟悉的程度是雙方根本可以稱得上是「朋友」的關係或熟識多年的客戶等。若是剛開始連絡的關係，建議您還是依照一般的尊稱用法。

(十五) Dearest Chris「給最親愛的克里斯」：

　　適用在非常親密的朋友、親人、伴侶之間的用語，很少人會使用在商業場合，否則便有過於拉攏關係、拍馬屁的嫌疑。

(十六) Hi Chris「嗨，克里斯」：

　　用"Hi"打招呼的方式，是與對方有一定的交情的情況，是一種非常隨性的稱呼方式，在某些正式的書信中並不建議使用。

「e-mail 商用書信」並不是都是嚴肅的，當你與對方已經有一定的熟悉程度的認識時，偶爾將嚴肅而沒有感情的稱謂改為用一種輕鬆而不失禮的方式表現也是被允許的。

在這裡順便一提一個有趣的名詞：

"Dear John Letter"，字面意思雖然是「給親愛的約翰一封信」，實際上是表示「絕交信」的意思。

三、 Body of letter「正文」

「正文」(Body of letter)是書寫這一封「e-mail 商用書信」的宗旨精神。"body"是「身體」之意，「信的身體」也就是「正文」之意。

一封好的商用書信，必須能夠發揮其應有的商業開發、溝通、解決問題的功能，這一部份也是收件人決定是否要保留或注意這封信的「關鍵時刻」。

商用書信最忌諱就是無關緊要的廢話，在撰寫商用書信時，你必須提出幾個問題先自我回答：

「內容的重點為何？」

「希望對方如何重視這個問題？」

「解決的方法為何？」

並不是用最艱澀、深奧的英文就表示自己的英文能力一流，因為如此一來，不但閱讀不易、也反而容易讓你的客戶產生你「很難相處」、「容易刁難」、「炫耀英文能力」的錯覺，因此，一封真正有效率的「e-mail 商用書信」是能夠用簡單的英文，就達到雙向溝通的功能。

　　此外，也切忌將商用書信當成閒聊的工具，把握"key word"「關鍵用詞」的使用，才是書寫一封成功的「e-mail 商用書信」的訣竅。

　　一般而言，不管是中文或英文的商用書信，在「正文」(Body of letter)中，都包括了三大「段落」(paragraph) 的內容：

　　(一) Opening「開場白」
　　(二) Middle「中間主文」
　　(三) Complimentary Close「結束語」

　　根據不同的商務訴求，每一個段落都必須扮演好各自的角色，才能夠有效地發揮「e-mail 商用書信」的溝通功能。以下就這三大段落做一個詳細的使用說明介紹：

(一) Opening「開場白」:

第一段的"Opening"（開場白）內容除了具備「簡單地向對方打招呼」的功能之外，也可以將你在之後的中間主文(Middle) 所要詳述的內容先做簡單說明，通常以簡單一兩句說明即可帶出重點。

一般而言，這一個段落可以有以下四種主題句型供您選擇使用：

---

1. 簡單打招呼：

句型 It's been a long time since ~「自從~後已經很久了」

例句 It's been a long time since we met in Hong Kong.

自從上一次我們在香港見面後，已經好長一段時間了。

句型 How is ~?「~好嗎？」

例句 How is your wife?

您的夫人好嗎？

---

2. 已知道對方的問題：

句型 We are terribly sorry for ~「關於~一事，我方感到很抱歉」

例句 We are terribly sorry for that accident.
關於那件意外，我方感到很抱歉。

句型 We regret to know ~「很抱歉得知~一事」

例句 We regret to know the death of your son.
很抱歉得知您愛子過世一事。

句型 We heard from ~「我方從~(某人/某處)得知」

例句 We heard that promotion form Mr. Smith.
我們從史密斯先生那兒得知此升遷一事。

句型 We heard of ~「我方得知~一事」

例句 We heard of your final decision.
我方得知您的最後決定。

3. 已收到對方的來信：

句型 Thank you very much for your letter dated ~

「感謝您於~（日期）的來信」

例句 Thank you very much for your letter dated 25th of January.

感謝您於一月廿五日的來信。

句型 We are pleased to receive your letter about~「很高興收到您關於~的來信」

例句 We are pleased to receive your letter about the price.

很高興收到您關於報價的來信。

句型 As requested in your letter, ~「根據您的來信要求~」

例句 As requested in your letter, we will consider his background.

根據您的來信要求，我們會考慮他的背景。

4. 回覆對方的詢問：

句型 In response to your letter, ~「在此回覆您的來信，~」

例句 In response to your letter, we suggest to you that we should meet.

　　回覆您的來信，我方建議雙方應該見面。

句型 As regards to ~「這事是關於~一事」

例句 As regards to your suggestion, we will reschedule this appointment.

　　這事是關於您的建議，我方將會更改這個約會。

句型 In connection with ~「這是有關於~一事」

例句 In connection with that offer, Mr. Smith will call you back.

　　這是有關於報價一事，史密斯先生會回您電話。

(二) Middle「中間主文」:

第二段的"Middle"(中間主文)內容就是這一封商用書信的「主題」(subject)，也是成就您的商業行銷策略是否能成功策馬入林的重要關鍵，在開場白簡單打過招呼後，您就應該盡快進入主題，千萬不要再提一些與主題無關緊要的旁枝末節的事件。

以下提供幾點書寫商業書信的重點建議，是您在撰寫此段書信的重要參考方式：

1. 重點(Point)：簡潔有力的文字敘述，避免過於冗長的說明。

2. 話題(Topic)：提供對方感興趣的話題。

3. 問題(Question)：提出問題以吸引對方注意。

4. 保證(Certification)：提供有利的證據、保證或輔佐資料。

以上的撰寫方式都是為了成功地達成商用書信在行銷時的助力，但是最終的目標，仍是以能成功引起對方的「注意」(Attention)，並發揮「實際行動」(Action) 的功效。

在各種商業行為中，不外乎「招商」、「報

價」、「交貨」、「付款」、「代理」、「合約」、「賠償」等議題，本書將在後續的各章節中逐一提供各種例句供您參考。

此外，在「e-mail商用書信」的行銷領域中，並不建議您附加(attach)任何的檔案，因為電腦世界的「病毒(Virus)」實在讓人不敢恭維，而一般人都被教育避免打開「附加檔案(Attachment File)」，所以建議您可以用「超連結」(hyperlink)的方式指引你的顧客「點選」(click)到相關「網頁」(website)。

此外，提供「免付費電話」(Toll-free)的訊息也是一個方式，但是根據網路行為者的調查研究統計，一般人寧願點選網頁查詢資料，而懶得拿起電話撥打給客服部門(CSR)。

Tip

❶ 「點選網頁」怎麼說？

"click"是「卡嗒聲」，聽起來是不是很像「點選」的聲音，所以「點選網頁」的英文就叫做"click website"。

❶ 撰寫e-mail英文商用書信的訣竅　❷ 常用商業例句　❸ 常用商業名詞　❹ 商業書信範例　❺ 商業常用詞彙

(三) **Complimentary Close**「客套結束語」：

　　一封成功的商用書信要能前後連貫，「主文」是從收信者的立場為書寫的觀點，而第三段的「客套結束語」則站在寫信者的立場結束，因此必須簡潔有力，通常只要一句話說明即可，具有「靜待佳音」、「靜待您的回覆」或「不勝感激」的意思，旨在傳達您的耐心等待及誠意。

　　以下是一些您可以利用的結束語的句型及例句，可以幫助您更快得到對方的回覆：

---

1. 期待語句：

句型　be looking forward to ~「靜待得到~」

例句　We are looking forward to receiving your reply.

　　我方期望得到您的回覆。

例句　I am looking forward to your comments.

　　我將靜待您的建議。

Ｔ Ｉ Ｐ

❷ look forward to「期待某事」的使用句型：

looking forward + $\begin{cases} 名詞 \\ 動名詞 \end{cases}$

"look forward to"（期待某事）的句型中，其中的 to 是介詞而非不定詞，所以後面不可加原形動詞。

2. 感謝語句：

句型 Thank you for ~「感謝您的~」

例句 Thank you for your cooperation with us in this matter.

感謝您對於此一事件給予我方的協助。

例句 Thank you again for your attention.

感謝您對於此事的注意。

3. 要求語句：

句型 Please inform[tell] us of ~「請告訴[通知]我方~（事）」

例句 Please inform us of your decision soon.

請盡快通知我方您的決定。

例句 Please tell us whether you may accept it.

請告知我方您是否會接受。

4. 詢問語句：

句型 Will you please ~?「能請您~？」

例句 Will you please reply us before this Wednesday?

能請您盡快在這個星期三前回覆嗎？

例句 Will you please send us a copy of your latest catalogue?

能請您寄給我方一份貴公司的最新目錄嗎？

5. 堅信語句：

句型 We trust [believe/await] ~「我方相信[堅信/等待]~」

例句 We trust you will now attend to this matter without further delay.

我方相信貴公司會毫不延誤地關心此事件。

例句 We await your satisfactory to our quotation [service/product].

我們等待貴公司對我們的報價[服務/商品]感到滿意！

6. 謙虛語句：

句型 You are welcome to ~「歡迎~」

例句 You are welcome to ask BCQ Company for any help.

歡迎您隨時要求 BCQ 公司任何的協助。

句型 You are the most welcome to ~「歡迎你做~(某事)」

例句 You are always the most welcome to contact us.

歡迎您隨時與我方聯絡。

7. 保持聯絡語句：

句型 Please don't hesitate to ~「請不要遲疑去做~(某事)」

例句 Please don't hesitate to contact us.

不要遲疑，請馬上與我們聯絡。

句型 Feel free to ~「不要客氣去做~(某事)」

例句 If you have any questions, feel free to call me.

如果您有任何問題，不要客氣打電話給我。

四、Terms of respect「敬辭」

　　有點類似中文「謹啟」、「敬啟」的意思，通常有以下幾種表現方式，由上至下依序為「最尊敬」到「普通程度」的敬辭使用：

Very sincerely yours,	「謹啟」	正式用法
Sincerely yours,	「謹啟」	
Faithfully yours,	「謹啟」	
Yours very truly,	「謹啟」	
Very respectfully yours,	「謹啟」	
Yours,	「謹啟」	
Best regards,	「謹啟」	非正式用法
Best wishes,	「謹啟」	
Warmest regards,	「謹啟」	
Regards,	「謹啟」	
Take care,	「保重」	
All the best,	「祝好」	
Be good,	「祝好」	
Cheers,	「祝愉快」	

　　「敬辭」視為一個句子，所以必須使用大寫的格式。

五、Signature「簽署」

從前只有紙張可以書寫的商用書信時，「簽署」這一部份就包括「寫信人」及「發信人」的簽名，如今 e-mail 盛行，「代為書寫文書」的情形根本不存在了，因此在此是以「發信人」的立場來解釋「簽署」的說明，主要有「發信人姓名」及「職銜」、「公司」三大部分。書寫的位置是以齊信件的左邊邊界為主。

雖然在信件的信頭可以看見發信人的身分或名字，但還是建議您能夠正式地告訴對方您的名字、職銜或公司名稱。以下幾種署名的方式，您可以自行選擇使用：

1. 只寫名字：

   Chris

2. 名字+姓氏：

   Chris Nietopski

3. 姓名+職稱：

   Chris Nietopski

   Assistant Manager

4. 姓名+職稱+公司名稱：

   Chris Nietopski

Assistant Manager

BCQ Co. Lit.

5. 姓名+職稱+公司名+聯絡方式

Chris Nietopski

Assistant Manager

BCQ Co. Lit.

Tel:886-2-86473663

chris@foreverbooks.com

六、Postscript「附註說明」

此即為大家所熟知的"P.S."，是針對當你完成一封商用書信後，補足其遺漏待說明的部分，或是當你必須中斷此文句的本文說明時使用，通常是在非正式信件中較常使用。"P.S."的全文是"Postscript"

但是現在有越來越多的情況是可以利用"P.S."作為加強印象的功能。例如：

P.S. I have to renew our quotation next Friday.

（附註，我下星期五要更新我們的報價單。）

此外，若是已經註明了"P.S."的說明文後，又要再補充說明，則可以用"P.P.S."再次說明。"P.P.S."的全文是"Post-Postscript"。

若是又要再補充第三個"P.S"呢？則可用"P.P.P.S."表示，並以此類推。

七、Attached file「隨信函 附加檔案」

1. 隨信函附上

句型 We will enclose ~ 「隨信函附上~」

例句 we will enclose our illustrated pictures and price list promptly.

我方將會立即隨信函附上我們的向量圖檔及報價單。

句型 Attached, please find ~ 「隨信函附上~」

例句 Attached, please find the final report.

隨信函附上最後的報告。

句型 We are enclosing a copy of ~ 「隨信函附上~的影本」

例句 We are enclosing a copy of the reply we sent you on April 11.

我們正隨信函附上我們四月十一日寄給您的信件的影本。

E- MAIL

1 撰寫 e-mail 英文商用書信的訣竅

句型 Attached you will find ~「隨信函附上你要的~」

例句 Attached, you will find a detailed agenda.

隨信函附上一份你要的議程細節。

句型 Enclosed, please find ~「附寄在內的是~」

例句 Enclosed, please find our calendar of events for the month of February.

附寄在內的是我們二月份的活動行程。

2. 郵件寄出

句型 We will send you ~「我們會把~寄給您。」

例句 We will send you the contract separately.

我們另外再把合約寄給您。

句型 We are pleased to send you ~「我們非常高興地寄~給你」

例句 We are pleased to send you the 2016 Summary Financial Report.

我們非常高興地寄給你二〇一六年財務總結報告。

2 常用商業例句
3 常用商業名詞
4 商業書信範例
5 商業常用詞彙

句型 Would you please send us ~?「您能將~寄給我們嗎？」

例句 Further to your request, would you please send us the new form first?

在此應您的要求，您能否寄給我方新的表格。

句型 We would appreciate it if you would send us ~「如果你能將~寄給我的話，我們將非常感激」

例句 We would appreciate it if you would send us the samples.

如果你能將樣品寄給我的話，我們將非常感感激。

Tip
① 網路語言的表情符號

網路語文中，逐漸發展出特有的「網路表情符號」，這是藉由許多簡單的符號所形成的許多種臉部表情，是一種非常通俗的心情表現，雖然不適合使用在正式場合的書信中，但是若你與對方相當熟悉，偶爾使用也是無傷大雅的。

^_^	高興
:-)	微笑
:-D	開心
\(^o^)/	舉手歡呼
(^O^)	大笑
*^_^*	可愛笑臉
;-)	使眼色
:-P	吐舌頭
:-O	張大口
@_@	疑惑、暈頭轉向
(ˇ-ˇ)	無耐
>_<	生氣

:-(	不悅
（"O"）	~~~生氣
⊙.⊙	睜大眼瞪你
→_→	懷疑的眼神
^(oo)^	豬頭
>"<‖‖	傷腦筋
~>_<~+	感人
::>_<::	哭泣
T_T	哭得很傷心
(><-)	痛
^_^;	尷尬
>_<‖‖	無奈尷尬
^_^‖‖	可愛尷尬
{{{(>_<)}}}	發抖
-_-b	流汗
..@_@‖‖‖..	頭昏眼花
(-_-)ZZZ	睡著了
(*_*)	慘了

follows:

An Export Company of rain wears in Taiwan is now making a business proposal for umbrellas which is said to have built a high reputation at home and abroad. Contact them by sending your e-mail to umbrella@yahoo.com.twAn Export Company of rain wears in Taiwan is now making a business proposal for umbrellas which is said to have built a high reputation at home and abroad. Contact them by sending your e-mail to umbrella@yahoo.com.twAn Export Company of rain wears in Taiwan is now making a business proposal for umbrellas which is said to have built a high reputation at home and abroad. Contact them by sending your e-mail to umbrella@yahoo.com.twAn Export Company of rain wears in Taiwan is now making a business proposal for umbrellas which is said to have built a high reputation at home and abroad. Contact them by sending your e-mail to umbrella@yahoo.com.tw

Chapter 2

常用商業例句

自我介紹

一、簡單開場白

例 We are desirous of extending our connections in your country.

我們擬拓展本公司在貴國的業務。

例 We have the pleasure of introducing ourselves to you as one of the most reputable exporters.

我們有這個榮幸向您介紹，敝公司是一家信譽優良的出口商。

例 Our company is well established and reliable.

我們公司有口皆碑且信用可靠。

例 We have full confidence that we will meet all your requirements.

我們有信心可以滿足您的所有需求。

常用商業例句

二、公司經歷

例 BCQ is an ISO, UL and NSF certified company.

BCQ是一家ISO、UL和NSF合格認證的公司。

例 We have been engaged in this business for the past 20 years.

敝公司從事這個業務已經有廿年的經驗。

例 Because of our past years' experience, we are well qualified to take care of your interests.

因為我們過去的經驗，我們具有極佳的勝任能力來照顧您的權益。

例 We have close business relations with the domestic private enterprises.

敝公司與國內私人企業有非常密切的商務關係。

例 We have four manufacturing plants in different countries: China, Japan, Spain and Philippine.

我們在不同的國家共有四個製造工廠：
中國大陸、日本、西班牙及菲律賓。

三、公司業務

例 We have been having a good sale of umbrellas and are desirous of expanding our market to your country.

我們的雨傘一直很暢銷，而我們想要在貴國擴展敝公司的經營市場。

例 We specialize in this line of business.

我們專門經營此項服務。

例 We specialize in this line over past twenty years.

我們專門經營此項服務已有廿年歷史了。

四、要求協助

例 We shall be much obliged if you will give us a list of some reliable business houses in Japan.

如果您能提供敝公司一些在日本具有可靠信譽的公司名單,敝公司將感激不盡。

例 We would appreciate it if you could kindly introduce us in your publication as follows.

如果您能在您的出版品上刊登敝公司以下說明,敝公司將感激不盡。

例 It will be great if you help me choose the right one.

若是您能幫助我選擇一個,那就太棒了!

例 We really need your help to make this event.

我們真的需要您的幫助以完成這件事。

Unit 02 會面

一、直接要求見面開會

例 I'd like to make an appointment to see you.

我想要跟您約個時間見面。

例 Would it be possible for us to talk to Mr. Black in person about that?

我們可否親自跟布萊克先生談此事？

例 Could we meet and discuss the matter in a little more detail?

我們可以見面再詳細討論一下這件事嗎？

例 Could we get together and discuss it a little more?

我們可否見面時再多討論一下？

例 I wonder if it would be possible for us to meet you at your office.

我在想，我們能不能在您的辦公室見個面？

例 Could I see Miss Jones sometime next week?

下個星期我能找個時間跟瓊斯小姐見個面嗎？

例 Would you arrange a formal appointment for us?

您能為我們安排一個正式的會面嗎？

例 I think we should meet as soon as possible.

我認為我們應該盡快見面。

例 I think we should meet someday.

我認為我們應該找一天見面。

例 If I should visit you then let me know.

假如我應該去拜訪你，就通知我。

例 We really need to discuss the plans in person.

我們真的需要當面討論計畫。

二、間接地要求開會

例 Mr. White would like to come and see you.

懷特先生想登門拜訪。

例 I'd like to meet Mr. Cruise at four o'clock in the afternoon.

下午四點鐘我想跟克魯斯先生見個面。

例 I'd like to see you tomorrow if you have time.

如果您有空，我想明天跟您見個面。

例 Should I visit you, or would you like to come over here and talk about it?

我能去拜訪您，或是您要過來討論呢？

例 How about come to my office?

您覺得來我的辦公室（見面）如何？

例 I would like to talk about it more often if you have time tomorrow.

假使您明天有空，我想要多討論一下。

例 What do you say if we talk about it at your office?

您覺得我們在您的辦公室討論如何？

例 I will have Debbie and Martin visit you and talk about it formally.

我會請黛比和馬汀拜訪您，並正式地討論這件事。

例 We have him change his mind.

我們要他改變他的心意。

例 I will have him call you back.

我會要他回你電話。

例 Is it convenient for you to call at my office?

如果您打電話到我的辦公室來，是否方便？

Tip

1 have someone do something

要求某人做某事

have 在此是帶有「要求」的意思。必須注意，在 have 之後，接受詞及原形動詞，千萬不能在動詞之前加不定詞 to。

【例句】

★ I firmly have him change his pants and clean up his mess.

我強烈要求他換褲子及清理這一切他所製造出來的髒亂。

★ Mr. Martin decided to have me check it out.

馬汀先生決定派我去檢查一下。

★ Now we have him make choices.

現在我們要求他做選擇。

2 常用商業例句

Unit 03 約定會面

一、詢問意見

例 What do you think?

您覺得呢？

例 What do you think about it?

您覺得這個如何？

例 What do you think about going to Japan?

您覺得去日本如何？

例 What do you think about the arrangement of BCQ company?

您覺得 BCQ 公司的安排如何？

例 When can we meet to talk?

什麼時候我們能見面談談？

例 When is it convenient for you?

您何時方便呢？

例 What time would be convenient for both of you?

您二位什麼時候方便呢？

例 I will arrange this meeting.

我會安排這次會議。

二、建議時間

例 How about four o'clock in the afternoon?

下午四點鐘如何？

例 I suggest that we should meet at two pm.

我建議我們下午兩點鐘見面。

例 What do you say five o'clock?

您覺得五點鐘如何？

例 How about ten o'clock on May 10th?

您覺得五月十日如何？

例 We would like to meet you at five pm.

我們想要五點鐘和您見面。

三、建議日期

例 We are going to visit you on the second of May.

我們將在五月二日拜訪您。

例 Shall we visit you in March when we pass through France?

當我們三月過境法國時,是否可以去拜訪您?

例 What do you think of next Friday?

您覺得下個星期五如何?

例 How about next Monday or the day after tomorrow?

下星期一或是後天如何?

例 How about meeting three days later?

大後天見面如何?

例 Could we make a tentative appoint-ment for Friday?

我們能暫時先約在星期五嗎？

四、建議時間及日期

例 How about the day after tomorrow at five o'clock in the afternoon?

後天下午五點鐘如何？

例 How about tomorrow night at seven o'clock? Is it OK with you?

明天晚上七點鐘如何？您可以嗎？

例 Are you free next Tuesday at two o'clock pm?

下個星期二下午兩點鐘您有空嗎？

例 Let's tentatively say next Wednesday at four o'clock.

我們暫時先約定下星期三下午四點鐘。

五、建議地點

例 How about four o'clock in my office?

四點鐘在我的辦公室如何？

例 How about ten am in my office?

早上十點鐘在我辦公室如何？

例 Come to my office tomorrow. We really need to talk.

明天到我的辦公室來，我們真的要好好聊一聊。

例 I really need to talk about it with you at my office.

我真的必須和你在我的辦公室討論一下。

❷ 常用商業例句

Tip

❷ What do you say ~?

你的意見是~？

字面意思雖為「你說什麼」，但其實更深一層的意思是「提出自己的建議後，再詢問對方的意見」，say 後面要加名詞或動名詞。

【例句】

★ What do you say Japan?

您覺得日本如何？

★ What do you say changing your mind?

您覺得改變您的主意如何？

Unit 04 確定會面

一、答應見面

例 See you soon.

再見。

例 Anytime you say.

什麼時候都可以。

例 I'm looking forward to meeting Mr. Jackson.

我期待和傑克森先生見面。

例 We are really looking forward to this appointment.

我們真的很期待這次的會面。

例 We'll be expecting you all.

我們期待各位的光臨。

例 We'll be waiting for Henry and Eileen.

我們恭候亨利和愛琳的光臨。

例 Mr. Jones and I are both expecting to meet you next week.

瓊斯先生和我兩人都很希望下週與您見面。

二、協調見面時間

例 What time is it convenient for you?

哪一個時間對您來說方便？

例 Which one do you prefer? Two or four o'clock?

兩點或四點鐘你比較喜歡哪一個？

例 Anytime between three and five.

三點鐘到五點鐘之間都可以。

例 I'm free after three o'clock.

三點鐘以後我有空。

例 I'll be out of town next Wednesday, but anytime after that would be fine with me.

下星期三我會出城去，在這之後我任何時間都可以。

三、決定見面時間

例 Ten o'clock is fine with me.

我十點鐘可以。

例 I'll see you at eleven.

那就十一點鐘見。

例 I think three o'clock is much better.

我覺得三點鐘比較好。

例 My boss and I will be there on 2 pm.

我老闆和我會在下午兩點鐘抵達。

例 I am writing to confirm the meeting arrangement.

來函確認會議的安排。

四、任何時間都可以見面

例 Please call on me anytime when it is convenient for you.

只要任何時候您方便,都歡迎來訪。

例 Please call on me it suits you.

您方便的時間隨時歡迎光臨。

例 Just give me a call before you come over.

您來之前打個電話給我就可以了。

例 Anytime you want I'll be there for you.

任何您想要的時間,我都會在那裡(等你)。

例 I am glad to discuss with you anytime you want.

我很願意在任何您想要的時間和您討論。

例 If you'd like to view our samples, you may come over anytime during regular business hours.

假使您想要看我們的樣品,只要在一般上班時間,您都可以親臨拜訪。

例 You can come over anytime you get ready.

等您準備好,隨時都可以過來。

例 As soon as you are ready to come over, just let me know.

一旦您準備好要過來,只要讓我知道就好。

例 You are always welcome to come over to the meeting.

永遠歡迎您來參加會議。

Unit 05 取消會面約定

一、無法會面的理由

例 I'm sorry, but I have an appointment with one of my clients tomorrow.

抱歉，我和我的一個客戶明天有約會。

例 I'm expecting some visitors from U.S.A. tomorrow morning.

明天早上我要接待一些來自美國的客人。

例 Carl said he is unable to be there.

卡爾說他無法去了。

例 Something urgent has happened. We won't be able to make it tomorrow.

有急事發生。明天我們去不成了。

例 Something happened, so I have to fly to Hong Kong this afternoon.

發生一些事，所以我今天下午要搭飛機去香港。

例 I will have to go to Hong Kong tomorrow morning.

我明天早上必須去香港。

例 I don't think it's a good idea. I couldn't make it.

我覺得這不是一個好點子。我可能辦不到。

例 I am afraid we have to cancel the meeting.

抱歉，我們必須取消會議。

例 I am writing to inform you Mr. White has decided to cancel the meeting..

謹來函通知您，懷特先生已經決定將會議取消。

二、會面將遲到

例 I am afraid it is a little late for me.

我覺得對我而言，恐怕有一點晚。

例 We may be a little late but please wait for us.

我們可能會晚一點到，但是請等我們。

例 Two o'clock? I'm not sure. I think it would be too late for us.

兩點鐘？我不確定。我覺得對我們來說太晚了。

例 This is to inform you that we are unable to be there on time.

來函通知您，我們無法準時到達。

三、取消會面

例 I'm afraid Mr. Jones has to cancel your appointment.

恐怕瓊斯先生必須取消和您的約會。

例 Louise and I have to cancel tomorrow's appointment.

路易斯和我必須取消明天的約會了。

例 Neil asked me to inform you that he is not going to visit you tomorrow.

尼爾要我通知你，他明天無法去拜訪你了。

例 I am writing to inform you Mr. Martin decided to cancel the appointment.

謹來函通知您，馬汀先生決定取消會議。

例 We decided to cancel the meeting today.

我方決定取消今天的會議。

例 My boss directed me to cancel the meeting.

我的老闆指示我取消會議。

例 For this reason, we have to cancel the meeting.

為了這個理由，我們必須取消會議。

例 We need to cancel the meeting originally scheduled for Wednesday, December 29 and reschedule for Monday, December 27.

我們必須取消原訂十二月廿九日星期三的會議，並（將會議）改為十二月廿七日星期一。

四、不確定是否能如期赴約

例 I'm afraid I would be busy all day tomorrow.

恐怕明天我會忙一整天。

例 I am afraid I will be busy all day long tomorrow.

我擔心我明天一整天都會很忙。

例 I am afraid I couldn't make it.

我恐怕無法到達。

例 I have to check my schedule.

我要查看我的行程。

例 I am not sure about it. Can I inform you later this week?

這件事我不確定，我可以本週晚一點通知你嗎？

例 Is it too late to cancel the meeting now?

現在取消會議會太晚嗎？

TIP
❸ make it

「成功完成某事」

"make it"表示「成功」、「完成某事」或「趕上」的意思。當有人邀請你時，若是你無法答應，你就可以回答：

"I am afraid I couldn't make it." (我恐怕無法成行)。

【例句】

A：Are you coming to the party?

你要參加宴會嗎？

B：I can't make it.

我不能參加。

Unit 06 更改會面行程

一、詢問更改會面的可能性

例 Would you like to decide on another time?

您要不要選定別的時間？

例 Could you change the schedule for me?

您能幫我更改時程嗎？

例 You did change the appointment, didn't you?

您有更改約會的時間吧？沒有嗎？

例 Would you like to reschedule the appointment?

您要重訂會面時間嗎？

例 How about Thursday at the same time?

星期四同一個時間好嗎？

例 How does Friday at the same time sound to you?

您覺得星期五同一個時間如何？

二、拒絕更改會面

例 I would rather not change the time.

我倒寧願不要更改時間。

例 I don't think it is a good idea to reschedule this meeting.

我覺得更改會議行程不是個好主意。

例 I don't think it is necessary.

我覺得不需要。

例 But I won't be available then.

但是屆時我不會有空。

例 I prefer the same day to next Friday.

比起下週五,我還是比較傾向同一天。

Unit 07 延後會面行程

一、建議延後會面時間

例 I'm afraid I'll have to postpone the appointment.

恐怕我必須把約會延期。

例 Can we make our appointment a little later?

我們的約會可以往後延些時間嗎?

例 I ask you to postpone tomorrow's appointment.

我要求您把明天的約會延期。

例 Would you please tell Andrew I have to postpone our meeting?

能請你告訴安德魯,我必須將我們的會面往後延期嗎?

二、拒絕延後會面時間

例 I prefer not to postpone the schedule.

我比較傾向不要延後行程。

例 I am afraid it is not convenient for me.

這恐怕對我不方便。

例 I won't be there if you postpone the schedule.

如果你延後會面,我可能無法參加。

例 It's impossible to postpone the meeting.

會議不可能延期。

例 Mr. White refuses to postpone the appointment.

懷特先生拒絕將會議延期。

Unit 08 提議

一、提出建議

例 May we suggest?

我方可以提出建議嗎？

例 We suggest that the following point should be added to our requirement.

我方建議，以下的重點應該要加到我們的需求中。

例 We suggest that you provide us with services.

我方建議，貴公司必須提供我們服務。

例 My proposal is that ~

我的提議是~

例 Here is what I propose ~

以下是我的提議~

例 I proposed Mr. White for the job.

我提議由懷特先生來執行這工作。

例 My proposal has two parts. They are ~

我的提議分為兩個部分,也就是~

例 He proposed a get-together this weekend.

他建議本週末聚會。

例 He proposed building a bridge across this river.

他建議在這條河上搭一座橋。

例 They propose to begin tonight.

他們建議今晚就開始。

例 There are a few points which I'd like to bring up concerning the contract.

關於合約,我想提出幾點看法。

例 I advised against their doing it.

我勸他們不要做這件事。

例 I advise you that you check your order form.

我建議您去確認您的訂單。

例 If you require prompt delivery, we suggest that you place your order before this Friday.

假使您需要立即的運送,我方建議,您要在這個星期五之前下訂單。

例 He suggested that those entries be blanked out.

他建議刪去那幾個項目。

例 How about having dinner with us this Friday?

本週五和我們一起用餐好嗎?

例 We planned to finfish the project on time.

我們計畫要如期完成企畫案。

二、要求對方提出建議

例 We should be glad if you would help us with your suggestions.

如果您能提供我們您的建議，我們將會非常高興。

例 I want your advice, sir. I don't know what to do.

先生，我需要您的指點。我不知該怎麼辦才好。

例 I need your advice about the member list.

我需要您有關成員名單的建議。

例 I need your advice about inviting certain people.

有關於邀請哪些特定人士，我需要您的建議。

例 What will you advise us that where we can hold the meeting?

對於我們要在哪裡舉行會議，您能給我們意見嗎？

例 Do you agree with me?

您同意我嗎？

Unit 09 代理權

一、要求代理

例 We are very interested in acting as your agent.

敝公司對擔任貴公司代理一事非常感興趣。

例 We would like to express our interest to be the supplier that you are looking for.

敝公司對成為貴公司正在找尋的供應商具有興趣。

例 We ask to be the exclusive agency for your computers in Taiwan.

敝公司請求作為貴公司的電腦在台灣的獨家代理商。

常用商業例句

例 We should be glad if you would consider our application to act as a sole agency for the sale of your products.

假使貴公司願意考慮敝公司所提成為貴公司商品銷售的總代理商，敝公司將非常欣喜。

例 We write to ask if you are interested in extending your export business to Taiwan by appointing us agent for the sale of your products.

敝公司來函欲了解，貴公司是否有意指定敝公司為銷售代理商，拓展出口商品至台灣銷售的業務。

例 Because you are not directly represented in Taiwan, we'd like to offer to be your agent.

因為貴公司在台灣還沒有直接的代理人，我們欲提供擔任貴公司在台灣的代理商業務。

二、代理權的優勢

例 We are sure that you know very well the advantages of representations.

敝公司確信貴公司非常了解代理商的好處。

例 We'd like to be your agent handling your export trade with Europe.

敝公司願做貴公司的代理，辦理貴公司對歐洲的出口貿易。

例 With many years' marketing experience, we offer our service as your agent in Canada.

我們在此行業中已有多年經驗，可以提供擔任貴公司在加拿大的代理服務。

例 With many good connections in the line, we write to act as your agent in Spain.

我們在此行業中已建立良好的關係，敝公司足以提供擔任貴公司在西班牙的代理業務。

三、商談代理過程

例 Thank you for your letter of September 25, in which you show your interest in being our supplier.

謝謝您九月廿五日來函表示有興趣成為敝公司的供應商。

例 Thank you for your letter of September 30, in which you indicate your desire to act as our sole agent for our products in Taiwan.

很感謝您九月卅日來信，説明貴公司欲成為敝公司商品在台灣的代理商。

例 We appreciate your request to act as our agent in your country, but we think it is premature for us to enter into agency agreement at the present stage.

感謝貴公司要求成為我方在貴國的代理，但我們認為現階段簽訂代理協議尚為時過早。

例 Regarding the question of agency, we should think it is not about the right time for us to consider it.

關於代理的問題，我們認為現在不是討論此問題的時機。

四、拒絕授權代理

例 Unfortunately, we have already signed a contract for a period of five years with another supplier in Taiwan.

不幸地,我們已經與另外一位在台灣的供應商簽訂了五年的合約。

例 We are obliged to decline your offer of sale agency proposal.

我們被迫拒絕您所提關於銷售代理的計畫。

例 We regret that we can't grant your request as we have already appointed an agent in your country.

很遺憾,敝公司不能答應貴公司的要求,因為敝公司已經在貴國指定了代理商。

例 We have decided to entrust BCQ with the exclusive agency for our products in Italy.

我們已經決定委託 BCQ 為我方商品在義大利的獨家代理。

五、答應授權代理

例 We have appointed you as our sole agent in Italy for a period of five years.

我們指定貴公司為我方在義大利為期五年的獨家代理商。

例 Regarding your proposal to represent us for the sale of our sporting clothes, we have decided to appoint you as our general agency in France.

關於貴公司建議銷售我方運動服飾，我們已決定委託貴公司為我方在法國的銷售總代理。

例 We have now made our decision to accept your proposal and to appoint you as our sole agent in Norway.

我們已經決定並接受貴公司的計畫，指定貴公司為我方在挪威的獨家代理商。

六、獲得代理權的肯定

例 We shall be much pleased to act as your sole agent in Sweden.

我們非常樂意能成為貴公司在瑞典的獨家代理商。

例 We are desirous of expanding the business and propose a sole agency agreement for your magazines for duration of ten years.

我們意欲擴大業務，今提出成為貴公司雜誌為期十年的獨家代理協議。

例 We can even act as your sole agent in marketing your existing products.

我方足以擔任貴公司在現有的商品領域的獨家代理商。

例 We welcome to cooperate with your company as your sole agent.

我方願意與貴公司合作，成為貴公司的獨家代理商。

TiP
❹ be interested in ~

對~有興趣

"be interested in~"是常見的片語,表示「對~感到極高的興趣」或「想要做某事」的意思。in 後面可以加名詞及動名詞。

【例句】

★ I am interested in fashion and food.

我對流行及食物有興趣。

★ We are interested in finding an English teacher for children.

我們想要替孩子們找一位英文老師。

★ We are interested in the Regional Office in Taiwan.

我們想要在台灣找一個地區性的辦公室。

Unit 10 詢價

一、初步詢問報價

例 We would like to make an inquiry.

我們想要詢價。

例 Please quote for us the prices of the items listed on the enclosed inquiry form.

請依所附需求表,為我方報價。

例 We have many inquiries for the under-mentioned goods.

我們收到許多下述貨品的詢價單。

例 I'm buying for chain stores in Australia. They are interested in Chinaware. I'd like to make an inquiry.

我是為澳洲的連鎖店採購商品,他們對瓷器非常感興趣。我想詢價。

例 We are interested in your products, please quote for us.

我方對貴公司的商品有興趣，請提供報價。

二、特定商品的詢價

例 A client of mine enquires for 100 cases of black tea bags.

我的一個客戶要詢價一百箱紅茶包。

例 We are desirous of your lowest quotations for hair driers.

我們想要貴公司頭髮吹風機的最低報價。

例 Please quote for us the latest prices for 25 containers of green tea.

請提供給我方廿五箱綠茶的最新報價。

例 Please send us your quotation for 200 suits of model SF4513.

請寄給我方貴公司型號 SF4513 商品 200 套的報價。

例 Please submit your quotation for the following goods:

請提供以下商品的報價：

例 We are interested in a couple of items in your new catalogue, and we'd like to know the prices.

我們對貴公司新目錄的一些商品有興趣，而且我們想要知道售價。

三、詢價的特定要求

例 We would like to know the price exclusive of tax of your computers.

我們想要知道你們的電腦不含稅的價格。

例 Please send us your best CIF quotation for micro wave ovens.

請報給我們微波爐最優惠的 CIF 價格。

例 Would you kindly please quote for us for your best offer with CIF Keelung?

能請貴公司提供 CIF 基隆的最優惠報價嗎？

例 Please quote us your best offer for the above inquiry based on FOB Keelung as well as CIF New York.

請提供以上產品報價，以 FOB 基隆和 CIF 紐約為報價條件。

例 Will you please send us a copy of catalogue, with details of the prices and terms of payment?

請寄給我方一份型錄，並註明價格和付款條件。

例 Will you please indicate details of discounts for regular purchases and large orders?

請提供我方定期採購與大量訂購的特價明細。

例 When quoting, please state terms of payment and time of delivery.

貴方報價時，請註明付款條件和交貨時間。

例 Please also indicate the delivery time in your quotation sheet.

請在報價單中註明交貨時間。

四、詢問報價的進度

例 The above inquiry was forwarded to you on January 15, but we haven't received your reply until now. Your early offer will be highly appreciated.

上述詢價已於一月十五日發給貴公司，可是我們到現在還沒收到貴方答覆，請早日報價，不勝感謝。

例 We haven't received your quotation until the date of May 10th.

我們直到五月十日都還未收到您的報價單。

ⓣⓘⓟ
❺ when+動名詞

當你要表達「某時間點做某事」時,可以利用"when+動名詞" 的句型當成從屬字句,之後再加上主要句子,可以既簡單又清楚地表達您的意思。

【例句】

★ Chris got engaged to her when traveling last winter.

去年冬天旅行時,克里斯與她訂婚。

★ Please send us an e-mail when finding the report.

當找到報告時,請用電子郵件通知我方。

E- MAIL

Unit 11 報價

一、感謝詢價

例 Thank you for your inquiry.

感謝貴方詢價。

例 We welcome your inquiry of July 30 and thank you for your interest in our products.

很高興收到你們七月卅日的詢價，並感謝你們對敝公司產品的興趣。

例 I take great pleasure in receiving your inquiry letter dated May 10, 2016 on the running shoes.

我很高興收到您二〇一六年五月十日有關運動鞋的詢價信件。

例 We would love to make an offer about sporting T-shirts.

我們很願意對運動 T 恤提供報價。

二、回覆詢價

例 Thank you for your interest in our products.

感謝你們對敝公司產品的興趣。

例 We have received your inquiry and will give you a quotation for 100 dozen of table lights as soon as possible.

我們已收到你方詢價，將儘快提供貴公司一百打的桌上型電燈的報價。

例 We have received your letter of March 6th about the quotation.

我們已經收到您三月六日有關報價的信件。

例 I take great pleasure in receiving your inquiry letter dated June 10th, 2016.

我很高興收到您二〇一六年六月十日的報價信。

例 Please tell us the quantity you require so that we can work out the offers.

請告訴我們貴方所需數量以便我方報價。

三、正式報價

例 We enclosed a copy of our price list.

隨信附上一份我們的價格清單。

例 Please find the attached information and our best offer for our products.

請參考附件資料和我們所提供最優惠的產品價格。

例 The heavy oil was quoted at US$17 per barrel.

重油報價為每桶十七美元。

例 We quoted 2,000 dollars for repairing the door.

我們要價兩仟元修繕那扇門。

例 We renew our offer of December 10th on the same terms and conditions.

我方基於原來條件，更新十二月十日之報價。

四、報價結語

例 Any increase or decrease in the freight after the date of sale shall be for the buyer's account.

出售日後運費如有上漲或下跌，均歸買方負擔。

例 All prices are subject to change without notice.

所有的報價隨若變更，恕不另行通知。

例 This quotation is subject to your reply reaching here on or before January of 20.

此報價以在一月廿日或一月廿日之前收到你的答覆才有效。

例 As the prices quoted are exceptionally low and likely to rise, we would advise you to accept the offer without delay.

由於所報價格特別低，並可能漲價，建議貴公司立即接受此報價。

例 We hope you will be satisfied with our samples and quotations.

我們希望貴公司能對我們的樣品和報價感到滿意！

例 This is a combined offer on all or none basis.

此為聯合報價，必須全部接受或全部不接受。

Unit 12 議價

一、價格過高

例 Your suggested price is rather on the high [low] side.

貴公司建議的價格偏高〔低〕。

例 I am afraid your price is above our limit.

您的價格恐怕超出了我們的界限。

例 I don't think we would make a bad purchase.

我不認為我們會用高價購買。

例 For all orders received by the end of the month, it is forty dollars each, and that is good on any size order.

只要月底前訂購，每片價格四十美元。不論訂量多少都是這個價格。

例 Your quotation is a little expensive.

貴公司的報價稍微昂貴。

二、拒絕接受報價

例 The price you offered is out of line with the market, so it is beyond what is acceptable to us.

您的報價與市場情況不相符，故我方無法接受。

例 We won't accept your quotation.

我們將無法接受您的報價。

例 I am afraid we won't be able to accept your quotation.

很抱歉，我們將無法接受貴公司的報價。

例 I am afraid that the prices you have quoted are much higher than our suppliers.

很抱歉，貴公司的報價較我們的供應商（的報價）還高。

三、價格比較

例 The Japanese quotation is much lower.

日本的報價就比較低。

例 The quotations we received from other sources are much lower.

我們從別處得到的報價要低得多。

例 We feel sorry that your price is more expensive than the BCQ's price for this article.

我們很遺憾發現貴公司的價格較 BCQ(公司)的價格來得高。

四、議價

例 Should you revise your prices, say, ten percent lower than the quoted one, we might persuade end users to conclude a deal.

假如你方能修改報價，比如降低 10 %，我們有可能說服客戶成交。

例 If you can make the prices a little easier, we shall probably be able to place an order.

如果您能再降一點價格,我們也許會下訂單。

例 Is it possible to shade the prices a little?

有沒有可能降一點價格?

例 We will place a trial order with you if the prices are competitive.

如果價格具競爭性,我們將會下試驗性訂單。

五、說服接受報價

例 It is the lowest price that we can offer you now.

這是現在我們所能提供給您最便宜的價格。

例 The price we offered has hit bottom and will soon pick up.

我們提供的報價已達到最低點,不久價格就會回升了。

例 The price is in line with the prevailing market and there is no room for reduction.

價格與目前市價一致，沒有降價的餘地了。

例 We may offer you a special discount of 10%.

我們可能提供你們九折的特別折扣。

例 We don't think we could cut our price to that extent as required.

我們無法把價格降到貴公司提出的限度。

Unit 13 樣品

一、要求提供樣品

例 A sample of the product may be required.

提供商品的樣品是必要的。

例 If available, please send us a piece of sample.

如果方便，請寄給我方一件樣品。

例 May I request some samples of your products before placing a formal purchase order?

我能在正式下單前，要求您提供商品的樣品嗎？

例 I would like to request some samples of your products before I purchase.

在我採購之前，我想要一些貴公司商品的樣品。

例 Please offer us two pieces of samples by the date of May first.

請在五月一日前提供兩件樣品給我方。

例 Would you please send us two samples of your products?

能請您提供貴公司兩件商品的樣品給我方嗎？

例 We would appreciate it if you would send us your samples as follows:

如下所示，如貴公司能寄出樣品給我方，我方將不勝感激：

二、願意提供樣品

例 We already sent you some free samples yesterday.

我方已於昨日寄給您一些免費樣品。

例 As requested in your letter dated October 12, we sent you the samples by air parcel on October 20.

依您在十月十二日的要求，我們在十月廿日用空運包裹寄給您樣品。

例 Four samples per each item were sent today via UPS Express.

我們已於今天將每款各四個樣品，以 UPS 快遞寄出。

例 Please give us your specific inquiries upon examination of the above as we presume they will be received favorably in your market.

在檢視過以上的樣品後，請告知您的特別需求，相信必能符合市場的需求。

例 We are sure that these samples will meet your requirements.

我們確信此樣品會符合您的需求。

例 Please let us know if our offer does not contain what you want in order to send you further samples.

假使我們的報價不包含您想知道的資訊,煩請告知,以便寄樣品給您。

三、樣品費用

例 Please let me know ahead the amount if payment is required for the samples.

若樣品要收費,請事先告知金額。

例 Can you mail us some free samples?

您能寄免費樣品給我方嗎?

例 We are planning to place a very large order. Can you mail us some free samples?

我們正打算下一筆大量訂單,您能寄給我方一些免費樣品嗎?

例 Would you please send us some free samples for us to try?

能請您提供我方一些免費樣品試用嗎？

例 Send us some free samples and I'll let you know which one we like best!

請寄給我們一些免費樣品，我會讓您知道我們最喜歡哪一款。

四、無法提供樣品

例 We are truly sorry that we can't send you samples.

非常抱歉，我方無法提供貴公司樣品。

例 We can't send free samples.

我方無法提供免費樣品。

例 You may place a sample order.

您可以下一張樣品的訂單。

例 As you can see our merchandise is high quality and expensive, we don't offer free samples.

如同您所瞭解的，我們的商品價值不斐，故無法提供免費樣品。

五、回覆收到樣品

例 Thank you for the sample.

感謝您的樣品。

例 Thank you for the samples, which you sent to us on Sep.10.

謝謝您九月十日寄給我們的樣品。

例 We have received the samples that you sent to us last week.

我方已收到您上週寄給我們的樣品。

例 We have received the sample which you sent us last Friday.

我們已經收到了上星期五貴公司寄來的樣品。

Unit 14 目錄

一、要求提供目錄

例 We would be grateful if you would kindly send us the catalouge of your lights.

如果貴公司能寄給我方您的燈具資料，我方將感激不盡。

例 We would appreciate it if you would send us your catalogue.

如貴公司能寄型錄給我方，我方將不勝感激。

例 Will you please e-mail us your latest catalogue by Friday?

能請您在週五前用電子郵件寄一份貴公司的最新目錄嗎？

E-MAIL

例 Please send us your catalogue, prices and sales conditions, if possible.

如果可能的話,請寄給我方貴公司的型錄、報價還有銷售條件。

二、答應提供目錄

例 Enclosed is our new catalogue.

附件是我們新的型錄。

例 We enclosed a copy of our latest catalogue, as requested in your letter of November 8.

依您十一月八日的來函要求,隨信附上我們最新的目錄。

例 Feel free to download our latest catalogue from here.

請直接從此處下載我方的目錄。

例 Our latest catalogue is now available to download with detail information on hundreds of products of every description.

我們最新的目錄現在可以下載,包含上百種商品的詳細說明。

例 Click here to go to the catalogue section.

請點選連接到目錄選項。

三、已收到目錄

例 Thank you for the catalogue you sent us last week.

感謝您於上週寄給我方的型錄。

例 We have received your catalogues, samples and price lists.

敝公司已收到貴公司的目錄、樣品和價格表。

例 I have received your full set catalogue. I am very interested in your products.

我已收到貴公司所提供的整份目錄。我對貴公司的產品非常感興趣。

例 Let me begin by thanking you for keeping me informed of your latest catalogue.

首先感謝您通知我有關貴公司的最新款型錄。

⑥ website

網站

當你的信中提及「可以至我們的網站」查詢相關資料時,比較專業及貼心的作法是,直接將您的網站網址複製在句子之後,讓對方可以直接連結網頁,而不必再至網頁瀏覽器連接至網站。

【例句】

★ You may browse our website http://foreverbooks.com.tw

您可以到我們的網站 http://foreverbooks.com.tw 瀏覽。

① 撰寫 e-mail 英文商用書信的訣竅　② 常用商業例句　③ 常用商業名詞　④ 商業書信範例　⑤ 商業常用辭彙

Unit 15 回覆

一、已收到來函

例 We have received your letter of March 3 about the quotation.

我們已經收到您三月三日有關報價的信件。

例 I take great pleasure in receiving your offer letter dated September 10, 2016.

我很高興收到您二〇一六年九月十日的報價。

例 We have received your letter of July 25.

我們已經收到您七月廿五日的信件。

例 Many thanks for your inquiry of Sep. 30 about our products.

感謝您九月卅日來函詢問我方的產品。

例 We are happy for your letter of Jan. 10 and are obliged to learn that you are interested in our products.

感謝您一月十日的來信，很高興知道你對我方的產品有興趣。

例 In reply to your letter of March 15, we enclose a copy of our revised catalogue.

回覆您三月十五日來函，我方附上一份我們更新的目錄。

例 All details are shown in our price list.

所有的細節都顯示在我方的價格明細中。

例 Thank you very much for sending us your quotation.

感謝您寄予我方的報價。

例 We are receipt of your letter dated September 14.

我方已收到貴公司九月十四日的來函。

① 撰寫 e-mail 英文商用書信的訣竅　② 常用商業例句　③ 常用商業名詞　④ 商業書信範例　⑤ 商業常用詞彙

二、要求回覆

例 Please answer promptly.

請立即回覆。

例 We are looking forward to your prompt reply.

我們期待您的立即回覆。

例 We are looking forward to hearing from you soon.

我們期待盡快得到您的回覆。

例 We look forward to receiving your reply in acknowledgement of this letter.

我們期望能收到您有關於此事的回覆。

例 Your prompt reply will be appreciated.

貴公司的立即回覆，我方將不勝感激。

例 We look forward to receiving your immediate reply and thank you.

我方期待貴公司盡快回覆，感謝您！

三、歡迎詢問

例 Please feel free to contact us if you have any questions.

若您有任何問題,請與我方連絡。

例 Feel free to contact me.

歡迎聯絡我方。

例 If you have any further questions, please feel free to contact us.

如果您有任何更進一步的問題,歡迎聯絡我方。

例 Please call me at 886-2-8647-3663 if you have any questions.

如果您有任何的問題,請打電話到886-2-8647-3663 給我。

例 I would like you to confirm this appointment by return.

我很樂意回覆確認此次的約會。

例 We appreciate your early confirmation.

我方很感激您的儘速確認。

例 I would appreciate having your reply by the 1st of April.

您能在四月一日前回覆，我方將不勝感激。

四、提出要求

例 Please let us know your decision immediately.

請立即讓我們知道您的決定。

例 Would you please tell us your suggestions within 10 days?

能請您讓我們在十天內知道您的建議嗎？

例 Please read the third paragraph of our letter No. 125 of August 5 so that you will get all the facts.

請閱讀敝公司八月五日編號 125 號信件的第三段內容，您就可以瞭解所有的事情。

例 Please inform us upon receipt of the quotation.

收到報價單時，敬請通知我方。

Unit 16 訂購前

一、訂購

例 We are going to make an order.

我們打算下訂單。

例 The Japanese are in the market for buying gold.

日本人正想購買黃金。

例 We've settled the terms for the contract in general, and we are going to place an order.

大體上我們已將合約條款都談妥了，接著我們就要下訂單了。

例 We are ready to place an order with you, but only one condition is that goods are confined to Taiwan.

我準備向你們訂貨，但是唯一的條件是，貨物只限賣給台灣的公司。

例 We are pleased to place the following orders with you if you can guarantee shipment from Taipei to Seattle by November 10.

若貴公司能保證在十一月十日前將貨物由台北運至西雅圖，則我方樂於向你訂購下列貨物。

例 We are ready to place a trial order.

我方已準備要下少量的訂單了。

例 We would like to place a trial order for the following items:

我方打算要下少量的訂單，如下列所示：

二、確認可以下訂單

例 Samples must arrive in Hong Kong before the beginning of December, otherwise the goods are useless.

樣品必須在十二月上旬之前送達香港，否則所訂貨物都將無效。

例 We are going to place an order. Please let us know what you can offer in this line as well as your sales terms, such as mode of payment, time of delivery, discount, etc.

我方即將下訂單。請告訴我方貴公司能提供什麼樣的產品，以及你們的銷售條件，如付款方式、交貨日期、折扣等。

三、回覆訂購

例 We have received your order No. 6451.

我方已經收到您編號 6451 的訂單。

例 We are ready to supply any quantity of your order.

我方準備好提供貴公司訂單所需之數量。

例 We thank you for your Order No. 1450 for 200 boxes of the glue.

感謝貴公司編號 1450 的 200 箱膠水的訂單。

例 Thank you for your order. We accept it and will dispatch the goods in the early of August.

謝謝您的訂單。我們接受此訂單,並將於八月初交貨。

例 We have received your order and have accepted your request to purchase.

我方已收到貴公司的訂單,並接受您的購買要求。

例 We will confirm that we have received your order and give you the expected delivery time.

我方將會確認已收到貴公司的訂單後,然後提供預計出貨的時間。

例 We regret to say that we are not in a position to accept your order.

很遺憾,站在我們的立場,將無法接受您的訂單。

四、說服訂購

例 We would advise your order without loss of time.

我們建議您應該不要猶豫，立即訂購。

例 Your orders have discontinued since the settlement of a claim you approved for the damaged goods of your order No. 9215.

自從通過貴公司索賠編號第 9215 號的訂單的損失後，貴公司的訂單就沒有再繼續了。

例 Once the order is confirmed, we need you to give us a pre-advice two weeks ahead you placing an order for we need to prepare the materials.

一旦訂單確認後，請在下訂單的前兩週預先通知我方，因為我們需要準備生產物料。

例 We therefore advise you to place order with us at an early date.

因此，我方希望貴公司能及早訂貨。

Unit 17 下訂單

一、下訂單

例 Thank you for the samples, which you sent to us on June 20. We are pleased to place an order as specified on the enclosed order sheet.

謝謝您六月廿日寄給我們的樣品。我們很樂意訂購隨信所附訂單詳列的貨品。

例 We have received your catalogue and price list, and now we order the following goods at the prices named.

已收到你方目錄和價格表，現按照所標示價格訂購下列貨物。

例 In reply to your letter of October 5 quoting us the prices of apples, we are pleased to place an order as mentioned in the enclosed sheet.

茲回覆你方十月五日蘋果的報價函，現如所附訂單，敝公司欲下訂單。

例 The following products have been selected from your price list; please supply them to us immediately.

從貴公司的價格表中，我方已選購以下的商品，請立即提供商品給我方。

例 Please execute the following order according to your quotation of Sep. 30.

請根據貴公司九月卅日所提供之報價，執行以下的訂單。

例 The quality of the Order No. 203 must be the same as your samples.

訂單編號 203 的商品品質必須和你們的樣品一樣。

例 Please confirm receipt of the Order No. 203.

請確認編號 203 的訂單。

例 If this order is satisfactorily filled, we will place large orders with you.

如果這次訂購令我方滿意，我方將會向貴公司下更大的訂單。

二、回覆已收到訂單

例 Thank you for your Order No.
1130-1.

謝謝您編號 1130-1 的訂單。

例 We are pleased to receive your
Order No. 1130-1 and confirm
acceptance of it.

很高興接到貴公司編號 1130-1 的訂單，並確
認予以接受。

例 Please notify us of the purchase
number.

請告知我們訂單的編號。

例 We shall fulfill your order by
November 30.

我們會在十一月卅日前完成您的訂單需求。

Unit *18* 運送

一、確認運送

例 Please inform us of the shipping date.

請通知我們裝船的日期。

例 Please ship the enclosed order immediately.

請立即安排所附訂單的出貨事宜。

例 Please tell us when our order will be shipped.

請告知我們的訂單何時會裝船。

例 Please inform us the shipping rate via e-mail by the end of December.

請在十二月底前,以電子郵件的方式,通知我方運送的費用。

例 We will inform you the shipping cost.

我方將會通知您運送的費用。

例 We will inform you the shipping status.

我方將會通知貴公司出貨狀態。

例 We will inform you the shipping charge once you finalize your order.

一旦貴公司完成訂購，我方將會通知貴公司運送費用。

例 Shipment will be arranged immediately after receipt of your order form.

在收到貴公司的訂單後，將會立即安排出貨。

二、通知已安排運送

例 We are pleased to inform you your order is now ready for shipment.

我方甚感榮幸通知您，您的訂單已經準備出貨了。

例 We are pleased to inform you that your order has been shipped.

我們在此通知您，您的訂單已經出貨了。

例 As the goods you ordered are now in stock, we will ship them without fail immediately.

因為貴公司訂購的商品尚有存貨，敝公司將立即安排出貨。

例 Your order for 300 doz. umbrellas will be shipped at the end of this month. You will receive them early next month.

您的三百打雨傘的訂單會在這個月底裝船出貨。您將會在下個月初收到這批貨。

三、客戶要求運送細節

例 Please do your best to ship our goods by S. S. "Hope".

請盡量以「希望」號輪船裝運我方貨物。

例 It is due to arrive at Hong Kong on September 30, and confirm by return that goods will be ready in time.

預計於九月卅日抵香港，請函覆確認，貨物將會及時備妥。

例 Something unexpected compels us to seek your cooperation by advancing shipment of the goods from October to July.

意外的情況迫使我們尋求你方配合，請將貨物裝運期由十月份提前到七月份。

四、廠商回覆運送細節

例 We have effected shipment under your L/C No. 123 issued by City Bank.

我們已經按照您在花旗銀行所開立編號 123 號信用狀規定裝船。

例 In consideration of your difficulties, we will leave the L/C as it stands and effect shipment without asking for amendment.

考慮到貴公司的困難，我們將照原信用狀規定裝船，不要求修改。

例 The shipment date is approaching. It would be advisable for you to open the L/C covering your order No. 2986 so as to enable us to effect shipment within the stipulated time limit.

船期即將來臨，建議你們開立訂單編號 2986 的信用狀，以便我方在規定的時間內裝船。

例 The shipping documents are to be delivered to you against payment only.

裝運單據將於付款後交至貴公司。

Unit 19 出貨

一、要求出貨

例 I would appreciate it very much if you could deliver the goods by this Friday.

如果你們能在這個星期五前出貨，敝公司將不勝感激。

例 We find both quality and prices of your products satisfactory and enclosed the order for prompt supply.

敝公司對貴公司的產品的品質和價格均感滿意，現寄上訂單，請供應現貨。

例 As the market is sluggish, please postpone the shipment of the order No. 203 goods to August.

由於市場疲軟，請將敝公司訂單編號 203 延遲至八月出貨。

二、回覆出貨

例 It will be around December when a new stock is supplied.

大概要到十二月左右才有新貨可供應。

例 I will let you know as soon as the new supply is available.

當有新貨時,我會盡快通知您。

例 In spite of our effort, we find it impossible to secure space for the shipment owing to the shortage of shipping space.

雖然敝公司已盡最大努力,卻因為船位不足而無法保證交期。

例 I'm sorry to say that the delivery is a week behind schedule.

很抱歉,交貨時間比計劃的行程晚了一個星期。

Unit 20 包裝

一、包裝指示

例 Let us have your instructions for packing and dispatch.

請告知包裝及裝運之所需指示。

例 This kind of packing costs more.

這種包裝的費用更多。

例 This kind of packing is much cheaper.

這種包裝比較便宜。

例 These are fragile.

這些是易碎品。

例 Most of them will be liable to go broken on arrival.

大多數到貨時容易破損。

二、包裝要求

例 Could you use cartons?

你們能不能用紙箱？

例 Could you use wooden cases; instead?

你們能不能改用木箱？

例 I am afraid the cartons are not strong enough.

我擔心紙箱不夠結實。

例 We suggest that you strengthen the cartons with double straps.

我方建議用兩根包裝帶紮固紙箱。

例 Enclosed please find the packing of other brands for your reference.

附件請詳見其他品牌的包裝方式，以供您參考。

例 A photograph of the split packing is shown in Figure 1.

分開包裝的圖片顯示在下方的圖形一。

❶ 撰寫 e-mail 英文商用書信的訣竅

❷ 常用商業例句

❸ 常用商業名詞

❹ 商業書信範例

❺ 商業常用語彙

例 I suggest that you find some packing boxes for my fragile dishes.

我建議您幫我易碎的盤子找一些包裝盒子。

一、要求商品說明

例 Would you compare your samples with the goods of other firms?

您可以將您的樣品和其他公司的產品作一個比較嗎？

例 Your glasses are very popular in Europe.

貴公司的眼鏡在歐洲很受歡迎。

例 I think we can strike a bargain with you if your goods are competitive.

我認為如果產品有競爭力，我們就可以達成交易。

例 We are pleased to find that your materials appear to be of good quality.

我們很高興發現，貴公司的材料品質相當優良。

二、回覆商品說明

例 Our model ER586 refrigerator is designed on modern lives and gives.

我們 ER586 型號的冰箱是為現代化生活及需求所設計。

例 So you will agree that it is the outstanding goods for economy.

所以您一定會同意這是市場上非常出色的商品。

例 We have none in stock now.

我們現貨中沒有此商品。

例 We are sorry that we do not have these goods.

很抱歉，我們沒有這些商品。

例 As our stocks are running short [low、out], we would advise your order soon.

因為我們的現貨逐漸短缺[減少、售完]，我方建議貴公司盡快下訂單。

例 We regret this article is out of stock.

很抱歉，此貨物已無存貨。

例 You can find that our products are of high quality.

您可以發現，我方的產品品質很好。

例 In respect to quality, I don't think that other fruits can compare with our fruits.

在品質方面，其它水果的品質很難和我們的水果相比。

Unit 22 付款

一、廠商要求付款

例 Attached is an invoice.

隨信附上發票。

例 We request your immediate payment.

我們要求您立即付款。

例 You should pay an invoice for US
$ 500.

根據發票，您應該付款五百美元。

例 We insist on payment in cash on
delivery without allowing any
discount.

我方公司堅持貨到付款，不打任何折扣。

例 You should make payment against
our documentary draft upon
presentation.

貴公司應憑敝公司的跟單匯票於見票時付款。

例 We insist on payment by irrevocable sight L/C.

我方堅持憑不可撤銷的即期信用狀付款。

例 Payment shall be made CWO by means of T/T or M/T.

訂單付款費用應以電匯或信匯的方式。

例 According to our Accounting Department records, your payment of NT$200,000 for your P/O#4512 is overdue since September 1.

根據我方會計部門的資料，貴公司訂單編號 4512 的廿萬元台幣的款項，自九月一日起便已逾期未繳納。

例 Our Accounting Department records indicated that US$800 had not been paid due to an oversight on your part.

我們的會計部門的記錄指出，美金八百元是因貴公司的疏忽而尚未結清。

二、客戶付款說明

例 We won't accept payment in cash on delivery, but may consider payment in cash with order.

我公司不接受貨到付款的支付方式，但可以考慮隨單付款的支付方式。

例 The amount concerned was forwarded to your account of the Taipei Bank by telegraphic transfer today.

上述的金額已於今日電匯至貴公司台北銀行的帳號。

例 Our terms of payment are by irrevocable L/C payable by sight draft against presentation of shipping documents.

我們的支付方式是用不可撤銷的信用狀支付，憑裝運單據見票付款。

❶ 撰寫 e-mail 英文商用書信的訣竅

❷ 常用商業例句

❸ 常用商業名詞

❹ 商業書信範例

❺ 商業常用詞彙

例 Payment by L/C will give the best protection to the exporters.

用信用狀付款為出口商提供了最好的保護。

例 Please extend the L/C to July 20.

請將信用狀的有效期限延至七月廿日。

例 We hope you will take actions to assist us with our financial difficulties.

希望你方採取措施，幫助我們克服資金困難。

例 We will not accept L/C 45 days. Please change it to L/C at sight.

我們無法接受見票四十五天後付款的信用狀條款，請將之修改為即期信用狀。

例 Please see the enclosed copy of our wire payment.

請參考我們附件的電匯繳款影本。

Unit 23 協調

一、提出協調要求

例 We would say that it was unwise of you to have done that.

我們只能說，您的所為是非常不明智的。

例 It seems to us that you ought to have done that.

似乎對我們而言，您應該做那件事。

例 We are afraid we cannot comply with your request.

我們恐怕無法順從您的要求。

例 If negotiation fails, it shall be settled by conciliation.

如果協商失敗，就需要調解。

❶ 撰寫 e-mail 英文商用書信的訣竅
❷ 常用商業例句
❸ 常用商業名詞
❹ 商業書信範例
❺ 商業常用辭彙

二、協調的方式

例 Shall I suggest that we meet each other half way?

讓我們各作一半的讓步吧。

例 We offer an opportunity of discussing a contract with you.

我們提供和您討論合約事宜的機會。

例 According to our arrangement only through the BCQ Company, can we export our products to American.

根據我們的計畫,唯有透過 BCQ 公司,我們才能出口我們的商品到美國。

例 I am so sure we can do something to help you.

我確定,我們能作些事來幫助貴公司。

例 The three sides reached an agreement to stop this project.

三方達成協議停止這項計畫案。

例 It's no problem for us to sign a contract with you.

對我們來說和貴公司簽約不成問題。

三、感謝答應協調

例 We should be glad to hear at your earliest convenience the terms and conditions on which you are prepared to supply.

能得知您盡早準備提供商品，我方將不勝感激。

例 I do appreciate the effort you're making towards concluding this transaction.

我很感謝你為達成這筆交易所做的努力。

例 Thank you again for your proposal and your understanding of our position will be appreciated.

感謝您的建議，對於您能夠體諒我們的立場，我們不勝感激。

⑦ 達成共識

「達成共識」是商業書信中最期望達到的結果，那麼「達成共識」的英文怎麼說？reach 是「達到」的意思，「達成共識」就叫做"reach an agreement"。

【例句】

★We don't reach an agreement on the financial perspectives.

在金融的看法上，我們沒有達成共識。

★We reach an agreement to carry out this plan.

我們達成要執行這項計畫的共識。

Unit 24 抱怨

【實用例句】

例 We have to complain to you about the damage in shipment, which has caused us so much trouble.

我們不得不向你方抱怨，裝運的破損給我方造成很大的麻煩。

例 We regret to inform you that the goods shipped per S.S. "Beauty" arrived in such an unsatisfactory condition.

我們遺憾地告知你方由「美女」號輪船運來的貨物令人十分不滿。

例 The importer has filed a complaint with our corporation about poor packing of the goods.

進口商為貨物的差勁包裝向我公司提出抱怨。

❶ 撰寫 e-mail 英文商用書信的訣竅

❷ 常用商業例句

❸ 常用商業名詞

❹ 商業書信範例

❺ 商業常用詞彙

例 We can assure you that such a thing will not happen again in future deliveries.

我們向貴公司保證，這樣的事件在以後的出貨中，將不會再度發生。

例 We regret to find out the total shortage of your goods is 5 pieces.

我方相當遺憾地發現，貴公司商品的短缺總額是五件。

例 I regret to say that I am unable to help you.

很抱歉，我愛莫能助。

例 I regret that you see it like that.

你這樣看待這件事，令我方感到遺憾。

Unit 25 賠償

一、提出賠償要求

例 As regards inferior quality of your goods, we claim a compensation of NT$100,000.

由於你方產品質量低劣，我方要求你方賠償十萬台幣。

例 We can't but lodge a claim against you.

我們不得不向你方提出索賠。

例 We hope you would compensate us for the loss.

希望你方賠償我方損失。

例 In order to solve the problem, we ask for compensation for the loss.

為了解決問題，我們為所受的損失要求賠償。

① 撰寫 e-mail 英文商用書信的訣竅

② 常用商業例句

③ 常用商業名詞

④ 商業書信範例

⑤ 商業常用剪輯

二、對賠償的回覆

例 We are sorry to learn that your goods were badly damaged during transit and the insurance company will compensate you for the losses according to the coverage arranged.

我們很遺憾得知貴公司貨物在運送途中嚴重受損，保險公司將按照投保險種賠償損失。

例 The claim is unfounded.

索賠的理由不充分。

例 We are not liable for the damage.

我方對損失沒有責任。

例 It is a case of force, which is beyond our control.

這是人力無法控制的「不可抗拒」事故。

Unit 26 道歉

一、致歉的原因

例 I am really sorry about that.

我真的為那件事感到很抱歉。

例 Will you forgive me?

您能原諒我嗎？

例 Will you forgave me for what I had said to you?

您願意原諒我對您說過的那些話嗎？

例 I owe you an apology for my rudeness yesterday.

我為昨天的粗暴行為向您道歉。

例 I apologize to you for not attending the meeting yesterday.

我因為沒有出席昨天的會議而向您表示歉意。

例 Please accept my apology, will you?

請原諒我好嗎？

例 We still feel sorry for the trouble that has caused you so much inconvenience.

我們仍然感到抱歉，對您造成這麼多不方便的困擾。

例 We reiterate that we will make every effort to avoid similar mistake in our future transactions.

我們重申我們會盡一切努力，去避免未來處理事務時發生類似的錯誤。

例 We regret to inform you that we are not in a position to enter into business relations with any firms.

我們很遺憾地通知您，我們無法和任何公司洽談商務事宜。

例 We are terribly sorry that we made so much trouble to you.

很抱歉，我們為貴公司製造了這麼多的麻煩。

例 Please accept our deepest apology for any inconvenience this matter has caused you.

請接受敝公司對此事所造成貴公司之不便的深深歉意。

例 I feel sorry that I am not able to make a corresponding move.

我為自己不能做相應的行動深感遺憾。

例 We hope this unfortunate incident will not affect the relationship between us.

我們希望這一不幸事件，將不會影響你我之間的關係。

例 We are sorry for the short delivery.

我們為交貨量短缺向貴公司致歉。

例 We are terribly sorry to learn from you that the quality of goods is not satisfactory to you.

我們為產品品質無法令貴公司滿意而感抱歉。

例 I feel sorry that this decision will affect business.

我甚感抱歉，這個決定將會影響生意。

二、因延遲回覆而表示歉意

例 We apologize for not replying to you earlier.

很抱歉未能儘早回信。

例 We are sorry for not answering your letter sooner.

未能及時回信給您，我們深表歉意。

例 We are sorry for not replying to your letter sooner.

未能及時回信給您，我們深表歉意。

例 We are sorry for not responding to your letter sooner.

未能及時回信給您，我們深表歉意。

例 Thank you for your patience.

感謝您的耐心。

1
7
3

Unit 27 結語

一、客戶式結語

例 We solicit your close cooperation with us in this matter.

我們懇求您對於此一事件能給予協助和合作。

例 Your courtesy will be appreciated, and we earnestly await your reply.

對於您的協助我們將感激不盡，敝公司將靜待您的回音。

例 Thank you for the trouble you've taken in this matter.

謝謝您為這件事費心了。

例 Thank you again for your proposal.

再次感謝您的企畫案。

例 We recommend this matter to your prompt attention.

我們建議您立即針對此事回覆。

① 撰寫e-mail英文商用書信的訣竅

② 常用商業例句

③ 常用商業名詞

④ 商業書信範例

⑤ 商業常用詞彙

例 Looking forward to your prompt action on this matter.

期待貴公司對於此事採取迅速的措施。

例 Thank you for your cooperation.

感謝您的合作。

例 Thank you for your kind reply on our invitation.

感謝您對我方的邀請的善意回覆。

二、廠商式結語

例 We are confident to give our customers the fullest satisfaction.

我們有信心能提供給我們的客戶最完整的滿意。

例 We hope to be of service to you and look forward to your comments.

以上希望能對您有所幫助,也靜待您的指教(意見)。

例 We are ready to be at your service and await your order.

感謝有這個榮幸為您服務,並靜待您的訂單。

例 Please call me any time if you have any questions.

如果您有任何問題,歡迎您隨時打電話給我。

例 We are looking forward to your reply.

我們期待您的回覆。

例 We are looking forward to your immediate answer.

我們期待您的立即答覆。

例 We should appreciate hearing from you immediately.

能立即知道您的消息,我們將不勝感激。

例 When an opportunity arises in the future, we'll contact you again.

將來有機會,我們將再與您取得聯繫。

例 Looking forward to entering into a business relationship with you.

期待與貴公司建立合作關係。

例 We look forward to the pleasure of hearing from you.

我們期待您的回音。

例 We look forward to a successful cooperation with IBM.

我方期待能與 IBM 公司成功的合作。

例 Please let us know if you agree to the terms.

如果貴公司同意此條款，請通知我方。

例 If there is anything remaining unclear, you are always the most welcome to contact us.

如果有任何不清楚的地方，歡迎隨時與我們聯絡。

例 If you have any questions, please do not hesitate to let me know.

若有任何問題，請不要猶豫讓我知道。

例 If you need any information, please contact us or visit our website.

如果您有任何問題,歡迎您與我們聯絡或瀏覽我們的網站。

例 We advise you to visit our website:
http://www.foreverbooks.com.tw

我方建議您瀏覽我方的網站:
http://www.foreverbooks.com.tw

Chapter 3

常用商業名詞

Unit 01 開發

　　商業行銷有許多種方法，一封成功的「開發信」是運用最廣的方式之一。商業開發信件的重點是要能夠引起對方的注意、激起對方的興趣並產生決定性的決策，為了刺激對方的需求，運用簡單明瞭、把握重點的寫作技巧是你必須要學習的訣竅：

1. look for	尋找～
2. be famous for	因～聞名
3. express our interest	我們對～感興趣
4. establish [enter into] relations	
	建立關係
5. take this opportunity	把握機會
6. visit our website	參觀我方網站
7. contact us	聯絡我方
8. be satisfactory with[to]	對～滿意
9. good business relationship	
	極佳的商務關係
10. high-quality	高品質
11. full-equipped	設備完善
12. the most reputable company	
	最具信譽的公司
13. be certified firm	具合格認證之公司

❶ 撰寫e-mail二英文商用書信的訣竅　❷ 常用商業例句　❸ 常用商業名詞　❹ 商業書信範例　❺ 商業常用詞彙

Unit 02 Order
訂單

「訂單」是商業上最能夠撼動人心的結果，當你發出一封商業開發信時，最期待的就是來自對方訂單的回覆，但是事情會不會常常事與願違呢？此時您就耐著性子好好找出問題的癥結。以下是常見有關訂單的常用名詞：

1. Order No. 109[P.O.#109	訂單編號 109
2. place[make] an order	下訂單
3. [initial]trial order	[初始]試驗性訂單
4. confirm[accept] an order	確認[接受]訂單
5. be on order	已在訂購
6. give an order for~	訂購~
7. execute[fulfill] an order	生產訂單
8. cancel an order	取消訂單
9. withhold the production	暫停生產
10. fulfill the obligation on a contract	
	履行合約
11. place an order with someone for~	
	向某人訂購~

E-MAIL

Unit 03 Inquiry
查詢

　　商務交易的第一步驟就是"inquiry" (查詢)，包括索取目錄、價格表、樣品等，以及細節如詢問功能、材質、成分等，必須有通盤在的了解才能有商業合作的機會，因為每種產業的商品性質不同，各產品的說明在此不贅述，以下是針對共同的、常用的查詢相關用法：

1. make an inquiry	詢價
2. credit enquiry	信用查詢
3. credit status	信用狀況
4. creditability	信用
5. financial standing	財務狀況
6. mode of doing business	交易方式
7. in detail	細節
8. products[productions] goods	產品
9. price list	價格明細
10. import[export] price	進[出]口價
11. sample	樣品
12. terms of payment	付款方式
13. time[date] of delivery	交貨時間[日期]

① 撰寫e-mail英文商用書信的訣竅 ② 常用商業例句 ③ 常用商業名詞 ④ 商業書信範例 ⑤ 商業常用詞彙

Unit 04　Offer
報價

「報價」是指賣方向買方提供商品價格的意思，內容相當廣泛，包括各種交易條件、雙方也可能會針對價格產生協調的動作，在此特別將常用的用詞列出供您參考：

1. catalogue	目錄
2. latest catalogue	最新目錄
3. quote someone	向某人報價
4. unit [selling][retail]price	單價[售價] [零售價]
5. counter offer	議價
6. term of payment	付款條件
7. delivery date	交貨日期
8. imported [exported] goods	進[出口]商品
9. quality	品質
10. quantity	數量
11. material	原料
12. quotation	報價單
13. the price exclusive[inclusive] of tax	不含稅[含稅]價

Unit 05 Packing

包裝

　　「包裝」的功能是能夠保護商品，讓商品在運送、搬運的過程中不會遭受任何的損壞，因此包裝商品的說明，是雙方在商用文書中，非常重要的一個溝通細節，您不得不重視。以下列舉包裝及出貨常用的名詞：

1. in strong wooden case with iron hopped	用鐵鍊箍住的木箱
2. in cardboard box	用硬紙板箱
3. in water-proof canvas	用防水帆布
4. in seaworthy packing	用耐航包裝
5. in regular export packing	用習慣出口包裝
6. in conventional export packing	用慣例出口包裝
7. no hook	請勿用鉤
8. keep cool[dry][flat]	保持低溫[乾燥][平放]
9. keep away from heat	遠離熱氣
10. fragile	易碎品
11. inflammable	易燃貨物
12. poison(小心)	有毒
13. glass with care	小心玻璃

Unit 06 Payment
付款

俗諺「親兄弟明算帳」，可見牽扯到金錢這件事是多麼地令人不得不小心處理。「付款」在所有商業行為中，常常會因為語焉不詳而造成一些誤解，因此在提及有關「付款」、「金額」的相關事項時，您不得不小心為之。常見的付款方式有以下幾種：

1. Cash with Order	訂貨付現(CWO)
2. Letter of Credit	信用狀付款(L/C)
3. Documents against Payment	
	付款交單(D/P)
4. Telegraphic Transfer Before Shipment	
	出貨前電匯(T/T)
5. Bank Cheque	銀行支票
6. Payment Before Shipment	
	出貨前付款
7. Cashier Cheque	銀行本票
8. Open Account	記帳(O/A)
9. open [establish] an L/C	
	開立信用狀
10. change [correct] an L/C	修改信用狀
11. L/C at sight	即期信用狀

_{Unit} 07 Insurance
保險

　　商品交貨過程中，因為長距離、長時間的運送過程，總會因為某些不可抵抗的災難造成損失，所以一般客戶都會為商品投保保險，以彌補萬一有災難發生時所造成的損傷，或是減少財務上的損失：

1. Marine Insurance	海上保險
2. Total Loss Only	全損險
3. Free from Particular Average	
	平安險
4. All Risks Insureance	全險
5. Wars Risk	兵險
6. Definite Insurance	確定保險
7. Provisional Insurance	臨時險
8. Theft Insurance	偷竊險
9. Warehouse to Warehouse Insurance	
	倉到倉保險

Unit 08 Agency
代理

　　「代理」包括代理行為及代理權，委託代理商依據一定的代理權，在某一境外特定區域代為銷售商品，並獲得利潤，這就是代理。此代理行為會牽涉到雙方的利益，不得不重視此一細節出現的相關語句：

1. agent	代理商
2. sole[exclusive] agent	獨家代理商
3. agency agreement	代理契約
4. return commission	佣金回饋
5. selling agent	銷售代理商
6. distributor	批發商
7. commission house	證券經紀商行
8. consignment sale	寄售
9. consignor、consignee	委託者、受託者
10. application for agency	代理申請

Unit 09 Regret & Thanks
道歉 & 感謝

在商業書信中，另一種常見的內容就是「道歉」、「抗議」、「抱怨」的問題。此時，如何利用文字表達自己的謙虛、歉意、不滿，或是安撫對方、彌補錯誤等，其中的分寸您必須拿捏得宜，別因為一時的「失言」而壞了大事：

1. be sorry for ~		為~道歉
2. thank you for ~		為~感謝
3. apologize for ~		為~道歉
4. appreciate		感謝
5. regret		遺憾
6. do our best		竭盡所能
7. tender our apology		致上我方的歉意
8. accept our deep apologies		
		接受我方道歉
9. understand our position		
		體諒我方立場
10. the inconvenience we caused		
		我方引起的不方便
11. make every effort		盡所有的努力
12. at your [our] end		在貴方[我方]

❶ 撰寫e-mail英文商用書信的訣竅　❷ 常用商業例句　❸ 常用商業名詞　❹ 商業書信範例　❺ 商業常用詞彙

Unit 10 Claim

索賠

　　在商業行為中，常常會因為人為疏忽或天然不可預估的外力造成商品的損壞、短缺等問題，若不是很嚴重事項，一般人往往是採取警告、道歉的處理方式即可，但若是嚴重到不可收拾的地步，則雙方會衍生出一些「索賠」的問題。以下就列舉一些常見造成索賠的原因及在面對「索賠」的問題時，經常使用的用語有以下幾種：

1. cancellation	解約
2. inferior quality	品質不良
3. different quality	品質不符
4. damage of the good	貨物毀損
5. delayed shipment	延遲交貨
6. claim、claim for the damage [loss]	
	申請賠償
7. complaint、make a claim [complaint]	
	提出索賠
8. meet the claim、accept the claim	
	接受索賠
9. adjust the claim、settle the claim	
	處理索賠
10. submit a claim	索賠訴諸仲裁

follows:

An Export Company of rain wears in Taiwan is now making a business proposal for umbrellas which is said to have built a high reputation at home and abroad. Contact them by sending your e-mail to umbrella@yahoo. com.twAn Export Company of rain wears in Taiwan is now making a business proposal for umbrellas which is said to have built a high reputation at home and abroad. Contact them by sending your e-mail to umbrella@yahoo. com.twAn Export Company of rain wears in Taiwan is now making a business proposal for umbrellas which is said to have built a high reputation at home and abroad. Contact them by sending your e-mail to umbrella@yahoo. com.twAn Export Company of rain wears in Taiwan is now making a business proposal for umbrellas which is said to have built a high reputation at home and abroad. Contact them by sending your e-mail to umbrella@yahoo. com.tw

Chapter
4

商業書信範例

Unit 01　請國外工會介紹客戶

We are desirous of extending our connections in your country.

我們擬拓展本公司在貴國的業務。

原文範例

Gentlemen,

We are **desirous** of **extending** our **connections** in your

country and shall be much obliged if you will give us a list of some reliable **business houses** in Los Angeles who are

interested in the **importation** of Chinese books.

We are an old and well-established

exporter of all kinds of Chinese books, especially Chinese **autobiography**.

Therefore, we are confident to give our **customers** the fullest satisfaction.

Your **courtesy** will be appreciated, and we **earnestly** await your prompt **reply**.

翻譯範例

敬啟者：

我們擬拓展本公司在貴國的業務，如果您能提供敝公司有關洛杉磯一些對進口中文書籍有興趣、可靠的公司名單，敝公司將感激不盡。

敝公司是一個具有多年經驗、信譽良好、進口多種中文出版品（特別是中文自傳）的企業，因此我們有信心能提供給我們的客戶最完整的滿意。

對於您的協助我們將感激不盡，敝公司將靜待您的立即回音。

關鍵單字

desirous	渴望、希望、急於希望~的
extend	(影響力、業務)擴展、擴大、(內容)充實
connection	商業關係、顧客、主顧
business houses	公司行號
importation	進口
autobiography	自傳、自傳的寫作
customer	客戶、顧客
courtesy	恩惠、優待、好意、幫助

earnestly	認真地、誠懇地、真切地、熱烈地
reply	回覆、回函、答覆

關鍵片語

be desirous of ~

想要~、急欲~、渴望

當你要表達「急欲作某事」時，"desirous" 是很好用的形容詞，"desirous" 代表「渴望」，是表達「希望從事~」的普遍用法。

【例句】

★ I am desirous of calling my parents.
我急欲要打電話給我的父母。

★ Both Sophia and Maria were desirous of finding a quick solution to the problem.
蘇菲亞和瑪莉亞二個人都希望找出能夠快速解決這個問題的方法。

★ We are desirous of seeing our lovely grandsons.
我們都急於見到我們可愛的孫子們。

Unit 02 請合作夥伴介紹客戶

We shall be obliged if you kindly introduce us to some of reliable importers.

貴公司如能將我方介紹給一些信譽良好的進口商，我方將不勝感激。

原文範例

Gentlemen,

We thank you for your **cooperation** for our business for the past five years.

Now we are desirous of enlarging our trade in **staple commodities**, but have had no good connections in Canada.

Therefore we shall **be obliged** if you could kindly **introduce** us to some of reliable **importers** in Canada who are interested in these lines of goods.

We **await** your immediate reply.

翻譯範例

敬啟者：

感謝貴公司過去五年來與我方的商務合作。

目前敝公司亟欲擴大我方主要商品的業務範圍，但是我方目前在加拿大並無良好關係。

因此若貴公司能將敝公司介紹給一些在加拿大對我方業務有興趣、信譽良好的公司，我方將不勝感激。

我們期待您的立即回覆。

關鍵單字

cooperation	合作
staple commodity	主要商品
be obliged	施惠、感謝
introduce	介紹
importer	進口商
await	等候

關鍵片語

We shall be obliged if ~

我方將不勝感激，假使~

這是一個地球村的世代，不但人無法離群獨居，企業也是如此，許多企業的發展也必須仰賴許多的關係才能達成。

當你希望某人或某家公司能幫你完成某事時，不論事件的大小，「感激你為我的付出」是必備的禮儀，因此，"we shall be obliged" 就很好用。

你也可以在這句話後面加上"if"，表示「如你幫我完成某事，我將不勝感激」的意思，表示再一次強調希望對方協助的事實在是很重要的。

"obliged" 也可以用 "happy"、"glad" 或是 "thankful" 表示。

【例句】

★ We shall be obliged if you will inform us of your needs.
   如果您能通知我方您的需求，我方將不勝感激。

★ I shall be obliged if you reply per return email.
   如果您能回覆每一封電子郵件，我將不勝感激。

★ I shall be obliged if you could please explain it more.
   如果您可以再多做解釋，我將不勝感激。

★ We shall be obliged if you could spare some valuable time for answering the following questions.
如果您可以撥冗回答下列問題，我方將不生感激。

★ We shall be obliged if you could call her for me.
如果你能為我打電話給她，我將不勝感激。

★ I am much obliged to you for the kindness.
非常感激你的好意。

★ I will be glad if you could approve this plan.
假使你同意這個計畫，我會非常高興。

Unit 03 公司自我介紹

We have the pleasure of introducing ourselves to you as one of the most reputable rain wears exporters.

我們有這個榮幸向您介紹，敝公司是一家信譽優良的雨具出口商。

原文範例

Dear Sirs,

We have the pleasure of introducing ourselves to you as one of the most **reputable** rain wears **exporters** in Taiwan, who has been **engaged** in this **line** of business since 1985; particularly we have been having **a good sale** of umbrellas and are desirous of **expanding** our **market** to your country.

We would appreciate it if you could kindly introduce us to the relative importers by announcing in your publication as follows:

"An Export Company of rain wears in Taiwan is now making a **business** proposal for umbrellas which is said to have built a high reputation **at home and abroad**. Contact them by sending your e-mail to umbrella@yahoo.com.tw"

We **solicit** your close cooperation with us in this matter.

翻譯範例

敬啟者：

我們有這個榮幸向您介紹，敝公司是台灣最具信賴度的雨具出口商之一，我們自從一九八五年即開始從事這個(雨具)行業；特別是我們的雨傘很暢銷，而我們想要在貴國擴展敝公司的經營市場。

如果您能在您的出版品上刊登以下的說明，將敝公司介紹給相關的進口商，敝公司將感激不盡：

「一家在台灣生產雨具信譽極佳的出口商，擬計畫發展雨傘相關的業務，若欲和其聯絡，請e-mail 至以下信箱：

umbrella@yahoo.com.tw」

我們懇求您對於此一事件能給於協助。

關鍵單字

reputable	有聲譽的、可尊敬、高尚的、規範的
exporter	出口商、輸出業者、出口國
engage	參與、贏得、吸引
例 engage in conversation 加入談話	
line	生產線
have a good sale	暢銷
expand	擴大、增大(業務、企業等)
例 expand a business 擴大企業	
market	行情、商品買賣、買賣的機會、需求、銷路、總銷售量
例 come into the market 上市	
例 lose one's market 失去買賣機會	
例 find a market for 為~找行銷管道	
business	事務、業務、工作、職業、行業、商店、企業、公司

at home and abroad	在國內外
solicit	請求、懇請、懇求、要求

例 solicit someone to do ~
懇求某人做某事

關鍵片語

We have the pleasure of introducing ourselves to you.

我們有這個榮幸向您介紹敝公司。

當你第一次與一個陌生的人聯絡時，「自我介紹」是少不了的手續，就是因為你是代表企業，更需要讓收信人一眼即明瞭這封 E-mail 的重點，因此第一句就是明白告知對方「我要介紹一下敝公司的背景」。

"pleasure" 表示「愉快、快樂、滿足、歡愉」的意思，"we have the pleasure of" 中文的解釋就是「我們有這個榮幸~（作某事）」的意思。

在上列句型中，of 也可以用 to 替代，後接原形動詞。

【例句】

★ We have the pleasure of reading her novel.
我們有這個榮幸讀她的小說。

★ May we have the pleasure of your company?
敬請出席。

★ May I have the pleasure to interview President George Bush?
我有這個榮幸訪問喬治‧布希總統嗎？

★ It is a pleasure to talk to her.
跟她說話是件很開心的事。

Unit 04 業務的介紹

We take this opportunity to place our name before you as being a buying, shipping and forward agent.

我們希望藉由這個機會向您介紹敝公司有關採購、船務及轉運的業務代理。

原文範例

Gentlemen,

In a **recent issue** of the "American Trade" from TIME magazine, we saw your name listed as **being interested in** making certain **purchases** in Taiwan.

We **take this opportunity to** place our name before you as being a buying, shipping and forward agent.

We have been engaged in this business for the past 20 years. We, therefore, feel that because of our past years' experience, we are well **qualified** to take care of your **interest**.

We **look forward to receiving** your reply in **acknowledgement** of this letter.

翻譯範例

敬啟者：

在「時代雜誌」最新一篇「美國商業」的報導中，我們看見貴公司名列在「有意在台灣採購商品」的名單中。

我們希望藉由這個機會向您介紹敝公司有關採購、船務及轉運的代理業務。

敝公司從事這個業務已經有廿年的經驗，由於我們過去的經驗，因此我們是具有極佳的勝任能力來照顧您的權益。

我們期望能收到您有關於此事的回覆。

關鍵單字

recent	最近
issue	報導
be interested in ~	對~有興趣
purchase	採購
take this opportunity	藉由這個機會
qualified	據必須條件的、合格的
interest	利益、利害關係、個人利益、權益
look forward to receiving	
	期望能收到
acknowledgement	回報、謝函、回執

關鍵片語

We are well qualified to ~
我們是具有極佳的勝任能力~(作某事)

當你想要表示「某人具有某一能力」或「足以勝任某工作」時，只要用 "be qualified to ~" 表示即可，是很適合「勝任」意思的片語。

"qualify" 是動詞，表示「取得資格」、「具備合格條件」、「合適」、「勝任」，通常用被動語句 "be qualified to ~" 的說法，表示「足以擔當~資格」的意思。"well" 是副詞，表示強調夠資格的語氣。

【例句】

★ We are well qualified to take care of your interest.
  我方夠資格負責您的權益。

★ We are well qualified to provide our clients with these services.
  我方夠資格提供我方的顧客這些服務。

★ We are well qualified to handle your business needs.
  我方夠資格處理您的業務需求。

★ My father is well qualified to be a teacher.
  我的父親具有當個老師的資格。

★ He is qualified to be a lawyer.
  他具有當律師的資格.

★ Mr. and Mrs. White are well qualified to take care of the homeless kids.
  對於照顧無家可歸的孩子，懷特夫婦勝任得很好。

Unit 05 表明對商品有興趣

At present we are interested in your goods.

敝公司目前對貴公司的商品極有興趣。

原文範例

Gentlemen,

We are one of the **largest** importers of **woolen** goods in Taiwan and shall be pleased to establish business **relationships** with you.

At **present** we are interested in your goods, details as our Enquiry Note No. 2564 **attached**, and are looking forward to receiving your **quotation as soon as possible**. When quoting, please **state** terms of **payment** and time of **delivery**. Your **prompt** reply will be much appreciated.

翻譯範例

敬啟者：

敝公司為台灣最大的毛線織品進口商之一，很高興和您建立商務關係。

敝公司目前對貴公司的商品極有興趣，細節請詳見所附 2564 號詢價單，敝公司期待能盡快收到貴公司的報價單，貴公司報價時，請說明付款條件和交貨時間。

感謝您的儘速回覆。

關鍵單字

largest	最大量的、最大的
woolen	毛料、呢絨、毛紡織物、毛線、毛料衣服
relationship	關係、關聯、感情關係
present	現在的、目前的、今天的、當前的
例 at present time	在現在
例 at present day	在今天
attached	連接的、附加的、附屬的
例 an attached file	附加的檔案
quotation	行情、時價、報價、估價
as soon as possible	盡快

| state | 陳述、說明、詳述 |

例 It is stated that ~
據說~、如上所述

| payment | 支付、付款、繳納 |

例 in payment for ~
支付~的、作為~的報酬

| delivery | (貨物、郵件)等的投遞、運送物品、一次投遞量、投遞 |

| prompt | 很快的、迅速的、快捷的、毫不拖延的、立即的 |

例 a prompt reply 迅速的回答

關鍵片語

be interested in ~

對~極有興趣

表示對某事、某人、物「有興趣」時,就叫做 "be interested in ~"。

"interest" 當動詞時,是表示「使產生興趣」,但是當句子是表示「對某事有興趣」時,通常為「被動式」語句的用法:"be interested in something"。

另外,"in" 後面可加「名詞」及「動詞」,要注意的是,若是加「動詞」時,要將「動詞」改為「動名詞」表示。

【例句】

★ I am interested in reading romantic novels.
我對讀愛情小說很有興趣。

★ My parents are interested in swimming in the morning.
我的父母對早泳有興趣。

★ Bob is interested in golf.
鮑伯對打高爾夫球有興趣。

★ We are interested in the following exhibitions.
我方對以下的展覽有興趣。

★ We are interested in designing security systems.
我方對設計安全系統有興趣。

★ We are interested in your article.
我方對貴公司的商品有興趣。

★ I am interested in your unique experience.
我對貴公司獨特的經驗感到有興趣。

Unit 06 要求提供商品型錄

Will you please send us a copy of your catalogue?

貴公司是否能寄給我方一份你們的型錄？

原文範例

Gentlemen,

We are very interested in your cameras and digital cameras. There is a steady and great **demand** in Taiwan for the **above commodities** of high quality at **moderate** prices.

Will you please send us a **copy** of your catalogue, with details of prices and items of payment? We should find it most helpful if you could also **supply** samples of these goods.

We are looking forward to hearing from you soon.

翻譯範例

敬啟者：

我們對貴公司的相機及數位相機非常有興趣。對上述所提商品，針對高品質、公道的價格標準，目前在台灣有穩定、大量的需求量。

貴公司是否能寄給我方一份你們的型錄，並附上價格、付款明細？如果貴公司能提供這些商品的樣品，將對敝公司有極大的幫助。

我們期待您的儘快回覆。

關鍵單字

demand	需要、需求

例 There is a great [a poor、little] de mand for this article.

這種商品需求量大 [不大、小]。

above	上述的(事情或物或人)
commodity	商品、必需品、物資、日用品

moderate
適度的、(大小等)中等的、一般的、(價格)適中的、公道的

例 moderate prices
公道的價格

| copy | 一本、一冊、一份 |

例 a copy of ~　　　一份~

例 Please send us 200 copies of the book.
請寄給我方200本此書。

| supply | 供給、供應(必需品等)、 |
| | 把~供應給、為~提供 |

例 supply with ~　　　提供~(某物)

關鍵片語

Will you please ~?

您是否能~?

　　當我們向對方提出某一項需求時，就非常適合使用 "Will you please ~?" 的句子，這是非常禮貌、客氣的請求用法，意思是指「您能否作某事？」、「能否請您作某事？」、「您是否願意作某事？」，不管是正式的書信或口語化溝通，都是一句既實用又得體的句子。

　　也可以將這句話改為 "Will you ~, please?" 的用法，意思也是一樣的。

【例句】

★ Will you please hold the door for me?
請你幫忙我扶著門好嗎？

★ Will you please do me a favor?
  你願意幫我一個忙嗎？

★ Will you tell her to give me a call,
  please?
  能請您轉告她回我電話嗎？

Unit 07　提供所需的樣品

We are glad to send you samples
of our goods you inquired.

我們很高興能寄給您一些您所要求的樣品。

原文範例

Dear Sirs,

Thank you very much for your letter dated
of December 20 about computers. We are
glad to send you two samples of our
goods you inquired. And the quotation is
as follow:

(1) Notebook: US$2,000 each, CIF New
York City

(2) PC: US$1,800 each, CIF New York City

Shipment will be made **within** four weeks
from **acceptance** of your **order**.

Thank you again for your interest in our
goods. We are looking forward to your
order soon and please **feel free** to **contact**
us if you have any questions.

翻譯範例

敬啟者：

感謝您十二月廿日有關電腦的來信。我們很高興能寄給您兩個您所要求的樣品。報價如下：

(1) 桌上型電腦　美金 2,000 元/台 CIF 紐約市

(2) 個人電腦　　美金 1,800 元/台 CIF 紐約市

在確認您的訂單之後的四個星期之內，會安排船運。

再次感謝您對本公司的商品感興趣。我們期待您立即的訂單，如果您有任何問題，歡迎您來電詢問。

關鍵單字

shipment	(船運、貨物)等裝運、裝運的貨物、載貨
within ~	不超過、在~的範圍內、在~以內

例 within a week
在一個星期內

例 within half a year
在半年之內

acceptance	接受、受理、贊成、答應
order	訂購、定單、定貨

例 place an order for an article
訂購物品

feel free　　　可隨意的

例 Please feel free to contact me.
歡迎與我聯絡。

---

contact　　　使接觸、與~聯繫、與~往來

例 contact someone
連絡某人

關鍵片語

We are glad to send you ~
我們很高興能寄給您~

人際關係是從互相用心開始的，當你要寄某樣物品給對方時，"be glad to send you ~" 是非常得體有用的句子，不僅表達出「我重視你」的意思，更有一種尊敬、愉悅的感覺在其中。

另外，也可以將 "glad" 改為 "happy" 或 "pleased" ，或將 "send" 改為其他相關的動詞，也有相同的意思。

【例句】

★ I am glad to send you my wishes.
我很高興能寄給你我的祝福。

★ We are happy to send them our wedding pictures.
我們很高興寄給他們我們的婚禮照片。

★ Chris is glad to send me the opening invitation.
克里斯很高興能寄開幕邀請函給我。

★ I am pleased having a friend like you.
我很高興擁有像你這樣的朋友。

Unit 08　商品的需求量

We are in a position to handle large quantities.

我方目前有大量的需求。

原文範例

Dear Sirs,

It has come to our attention through our Chinese friends in Taiwan that you are one of the **foremost** manufacturers and exporters of CD players in Taiwan.

Being in the market for CD players, we shall be greatly obliged if you will send us a copy of your catalogue, informing us of your best terms and lowest prices CIF Taiwan. As we are **in a position** to handle large **quantities**, we hope you will **make an effort to submit** us really **competitive** prices.

We should appreciate hearing from you immediately.

翻譯範例

敬啟者：

透過我們在台灣的朋友，我方了解到貴公司為台灣 CD 播放機主要的生產及輸出公司。

身處在這個 CD 播放機的行業中，假使貴公司能寄一份貴公司的型錄，並告知貴公司銷售最佳商品 CIF 到台灣的最低價格，我們將不勝感激。因為我方目前需求量頗大，敝公司希望貴公司能盡量提供我方具有競爭力的價格。

能立即聽到您的消息，我們將不勝感激。

關鍵單字

foremost	最先的、最前的、第一位的、一流的、主要的
in a position	處於~地位
quantity	大量、許多
例 a large [small] quantities of wine 大量[少量]的酒	
make an effort to ~	盡力去做某事
例 submit his proposal 呈交他的計畫	
competitive	競爭的、好競爭的
例 competitive prices	具競爭力的價格

關鍵片語

As we are in a position to ~
因為我方所處的地位~

當你要說明身處在「某一個情境」、「某一個地位」、「某一個身分」時，就可以用 "in a position" 表示，這句話可以為你目前的某種行為作為解釋，以表明為何自己有如此的行為或是結果，是一句非常正式的商業書信用語，一般而言，口語化英文中較少使用。

慣常使用的片語為 " be in a position to do sth"，通常表示因為有經驗、有錢或有能力可以做某式的意思。

【例句】

★ They are in a position to bargain.
　他們可以討價還價。

★ I am in a position to be a good daughter.
　我身為一個好女兒的身分。

★ Mary and John were in a position to keep the secrets for their company.
　身為公司的員工，瑪莉和約翰要保守秘密。

★ They are in a position to be parents.
他們是身為父母的身分。

★ We are now in a position to be able to understand.
我方現正處於能夠理解的狀態。

★ This course puts me in a position to get a senior management job in tourism.
這個課程足以幫助我得到在旅遊業更高級的管理技巧。

Unit 09 商品的數量

We are ready to deliver any quantity of table lights from stock.

我們準備好可以從現貨中運送任何數量的桌上型燈具。

原文範例

Dear Sirs,

We appreciate your inquiry of May 25 about our goods.

As requested in your letter of table lights, we are glad to **enclose** a copy of our new catalogue in which you will find many items that will **interest** you.

All **details** are shown in our **price list**. We are ready to **deliver** any quantity of table lights **from stock**.

We are pleased to quote you as follows and can **promise** shipment within 20 days on **receipt** of your L/C.

All prices **are subject to** change **without notice**.

翻譯範例

敬啟者：

感謝您五月廿五日對我們產品的詢價單。

關於您信件中對於桌上型燈具的詢價，隨信附上一份我們新的目錄，您可以找到您有興趣的商品。

所有的細節都在我們的價格表中。我們準備好可以從現貨中運送任何數量的商品。

我們很高興給您我方的報價，我方保證收到您 L/C 收據後，會在廿天內安排船運。

所有的報價隨時更動，恕不另行通知。

關鍵單字

enclose	使封入、附上、裝入

例 enclose a picture with a letter
隨函附上照片一張

例 Enclosed please find a check for 2,000 dollars.
茲附上兩千元支票一張。

interest	使產生興趣、使關心
detail	詳細狀況、細節
price list	價格表
deliver	運送貨物

224

from stock	由現貨中
promise	允諾、答應
receipt	收到、接到、發票、收據
be subject to ~	必須得到~的、需要~的

例 The plan is subject to your approval.
該計畫須經您的批准。

without	不曾、沒有、毋需

例 enter a room without knocking
沒有敲門就走進房間

notice	公告、告示、佈告、招貼、啟事

關鍵片語

be ready to do something
準備去作某事

　　當你花了一番功夫準備去做某事時，就可以用 "be ready to do ~" 的句型。這個句型不但適合在商業書信中使用，在一般口語化英文的使用頻率也相當頻繁。

　　此外，當別人詢問你："Are you ready to go?"（準備好要走了嗎？），你也可以只要簡單地回答 "I am ready."（我準備好了）。

【例句】

★ I am ready to go to Japan with my parents.
我準備要和我父母去日本。

★ Jeff is ready to propose to her.
傑夫準備向她求婚。

★ Mr. and Mrs. Whites are ready to buy that house.
懷特夫婦準備要買那棟房子。

★ A：Come on, let's go.
快一點，我們走吧!

B：But I am not ready.
可是我還沒準備好。

Unit 10 商品報價

We quote some of them without engagement as follows:

我們主動提供一些非合約性商品的報價，如以下所列：

原文範例

Dear Sirs,

As we are given to **understand** that you are interested in the Taiwan sporting goods, we take this opportunity of introducing ourselves as a reliable trading firm, established 10 years ago and dealing in the sporting ever since with fair **record** especially with the Southeast African countries.

In order to give you an idea, we quote some of them without engagement as follows:

(1) Bags, cotton & fibers mixed

(2) White, US$20 per doz. CIF your port

We are sending you some free samples.

Please give us your **specific inquiries** upon **examination** of the above as we **presume** they will be received favorably in your **market**.

We hope to be of service to you and look forward to your **comments**.

翻譯範例

敬啟者：

敝公司獲悉貴公司對於台灣的運動用品很有興趣，我們藉此機會向貴公司自我介紹，本公司為信用可靠的企業，我們經營運動商品有十年的時間，自從和南非國家合作以來，也都擁有良好的成績。

為了讓貴公司有一些概念，我們主動提供一些非合約性商品的報價，如以下所列：

(1) 提袋、棉和纖維混紡

(2) 白色，美金 20 元/打，交貨條件：CIF 至您的港口

隨信附上一些免費樣品，在檢視過以上的樣品後，請告知您的特別需求，相信必能符合您市場的需求。

希望能為您服務，也靜待您的建議。

關鍵單字

understand	懂得、理解、有理解力、聽說

例 The situation is better, so I understand.
我聽說形勢有好轉。

record	記錄、登錄、成績
in order to	為了要做某事
specific	特定的、特有的、獨特的、(陳述等)明確的、具體的
inquiry	詢問、探問、探聽、調查、追求、查問
examination	調查、檢查、審查
presume	假定、推測

例 Mr. Jones, I presume?
(碰見面熟的人說)您就是瓊斯先生吧？

market	商品行銷地區
comment	對時事問題等的評論、評述、批評、註解

例 No comment.
無可奉告。(被採訪者不願回答任何問題時的一種回答方式)

關鍵片語

Something as follows:

某事如以下所說明：

"follow" 為「跟隨」、「追隨」的意思，
"as follows" 代表「如下」、「如此」的意思。

當你說明某事時，其細節問題請對方「詳見以下說明」時，就叫做 "as follows"，因此通常在這句話後面一定會「追隨」有其他更詳細的說明或解釋，以補充前面所表明的「詳見如後說明」的意思，千萬不要將此句當成最後一句使用，否則閱讀者可能會覺得莫名其妙：「明明說『詳見以下的說明』，怎麼不見後面更詳細的說明？」

此外，要特別注意，"as follows" 是一個標準用法，不可因為後方所接的說明事項是複數而更改為 "as follow"。

【例句】

★ I answered as follows:
我的答覆如下：

★ You will see the results as follows:
您會看到以下的結果：

★ We have made a decision as follows:
我們已經做了以下的決定了：

★ My requests are as follows:
我的需求如下：

★ The only exceptions allowed are as follows:
唯一的例外如下：

★ Our main conclusions are as follows:
我方的最主要的結論如下：

Unit 11 報價詢問

Is there something wrong with our prices?

是否我方的價格有問題？

原文範例

Dear Sirs,

On March 28 we received a request from you for some prices and samples. We replied to your letter immediately on March 29, and when we did not hear from you, we wrote again to you on April 8.

Is there **something wrong** with our prices? Or you are not **satisfied** with our samples. It is sincerely hoped that you will take **advantage** of this opportunity and **favor** us with your order **without** loss of time.

We wish you have been satisfied with our **service** and your reply will be very appreciated.

E-MAIL
②
③
③
①
撰寫 e-mail 二英文商用書信的訣竅
②
常用商業例句
③
常用商業名詞
④
商業書信範例
⑤
商業常用詞彙

翻譯範例

　　敬啟者：

　　我們在三月廿八日收到貴公司有關商品的詢價及樣品需求，我方已立即於三月廿九日回覆。目前我方尚未得知貴公司的回覆，因此我方再度於四月八日來函詢問。

　　是否為我方價格的問題？或是您不滿意我方的樣品？我方竭誠希望貴公司能好好把握此機會，您的盡速下訂單將讓我方不勝感激。

　　我方希望貴公司能滿意我方的服務，而對於您的盡速回覆，我方將不勝感激。

關鍵單字

something	某事、幾分、有點
wrong	錯誤、有問題
satisfied	滿意的
advantage	有利條件、有利因素
favor	賜與、贈與、給予 (榮譽等)
without	無~、沒有~、缺少~
service	服務、供職、效勞

關鍵片語

Is there something wrong with ~?
是否為~的問題？

當你滿懷疑問卻一直無法獲得解答時，應該如何發問？

商業行為中，因為對方遲遲沒有回應的確會讓人坐立難安，此時你就需要主動出擊。利用 "Is there something wrong with~?" 的句型（~有問題嗎？）就是一個找出問題的方式，能讓對方深刻感受到你的積極與誠意。

【例句】

★ Is there something wrong with our proposal?
是否是我們的計劃有問題？

★ Is there something wrong with you?
你是否有問題？[你還好吧？]

★ Is there something wrong with our services?
我們的服務有任何問題嗎？

★ Is there something wrong with the shipment?
出貨有任何問題嗎？

★ Is there something wrong with Maria? She looks so upset.
瑪莉亞還好吧？她看起來很沮喪。

So.d...

Unit 12 延誤詢價

We are very sorry for the delay for your inquiry on October 25.

很抱歉延誤貴公司於十月廿五日所提之詢價。

原文範例

Dear Sirs,

We are very sorry for the delay for your inquiry on October 25. Enclosed is our Prices List No. 200310-14.

Four samples **per** each item were sent today. As for the pricing for the requested items, please have a **look** at the attached quotation.

This quotation is **subject** to your reply reaching here on or before November of 10. We hope to **receive** your orders by return **within** 10 days.

Please **confirm receipt** of all the samples and quotation by return. Thank you very much!

翻譯範例

敬啟者：

很抱歉延誤貴公司於十月廿五日所提之詢價，附件為我方編號 200310-14 之價格明細清單。

我們已於今天將每款各四個樣品寄出，至於貴公司對所要求的品項報價部分，請參考附件的報價單。

此報價以你的答覆在十一月十日之前到達這裡才有效。

在收到樣品和報價後，請回覆並確認。謝謝！

關鍵單字

per	每、每一
例 \$5 per man	每人五元

look	注意、查明
例 have a look at	察明、檢視

subject	必須得到~的、需要~的
例 The prices are subject to change.
價格可能有變動。
例 The plan is subject to your approval.
該計畫須經您的批准。

receive	收到、接到、得到

within	不超過、在~的範圍內、在~以內
confirm	一步確定、再次確定、加強(決心等)、使更堅定
receipt	發票、收據

關鍵片語

be sorry for the delay for ~
很抱歉延誤~

　　當你需要在商業書信中表達歉意時，誠意及迅速回應是兩大關鍵點。如何能夠兼顧這兩點呢？"be very sorry for ~" 是非常實用的片語，若是必須要道歉的主題為「延遲」，則只要再加上 "delay"，就可以成功的表達你衷心的歉意。

　　另外，若是 "sorry" 這個普通的道歉還不足以表現你致歉的誠意，也可以改用 "terribly sorry" 替代，表示「非常地抱歉」。

【例句】

★ I am very sorry for the delay for my homework.
很抱歉遲交我的作業。

★ We are sorry for the delay for the invitation.
很抱歉延誤發送邀請函。

★ She is terribly sorry for the delay for her help.
對於她遲來的幫助，她感到很抱歉。

Unit 13 回覆對方的來信

We have received with many thanks your letter of May 20.

感謝您五月廿日的來信。

原文範例

Dear Sirs,

We have received with many thanks your letter of May 20, and wish to express our sincere gratitude for your kindness in publishing our wish in your "Trade Opportunity".

We believe the arrangement you kindly made for us will connect us with some **prospective buyers** and bring a satisfactory result before **long**.

We thank you again for you taking **trouble** and wish to **reciprocate** your **courtesy** sometime in the **future**.

翻譯範例

敬啟者：

感謝您五月廿日的來信，對於您的「商務機會」刊物中刊登我方的訊息，表達我方誠摯的謝意。

我們相信您為我方善意的安排，將會讓我方獲得一些有機會的買家，相信不久之後會有令人滿意的結果。

再次感謝您撥冗協助，希望將來有機會報答您的協助。

關鍵單字

prospective	未來的、將來的、有希望的、預期的
buyer	買家、購買者
long	以上的、長達～

例 for long
長久(用於疑問句、否定句、條件句中)

例 Will he be away for long?
他將外出很長一段時間嗎？

trouble	煩惱、憂慮、苦惱
reciprocate	報答、回報
courtesy	恩惠、優待、好意、幫助

| future | 未來、將來、今後 |
| 例 in the future | 在將來 |

關鍵片語

We have received with many thanks ~

我方已收到 ~並感謝您 ~

　　當你收到某人的來信，回信給對方時就可以用 "We have received with many thanks your letter"。

　　"with many thanks" 表示「充滿謝意」的意思，是一種極為正式的書信用法，表示「感謝您作某事」的意思，口語化英文中倒是很少會這麼使用。

　　"We have received with many thank" 也不一定只能是回覆對方來信，也可以表示收到對方禮物或幫助時的一種表達謝意的方式。

【例句】

★ I received with many thanks your reply.
　已收到您的回覆，不勝感激。

★ I received with many thanks your responses.
　已收到您的回應，不勝感激。

★ We have received with many thanks your letter of September 30.
我們很感激收到您九月卅日的來信。

★ I have received with many thanks your payment.
我很感激收到您的貨款。

★ We have received with many thanks your presents.
我們很感激收到您的禮物。

★ All donations are received with many thanks.
所有的捐獻均已收訖，不勝感激。

Unit 14 回覆議價

Up to now we can not entertain your counteroffer.

我們至今無法接受你們的議價。

原文範例

Dear David,

We appreciate your counteroffer of August 10 about our goods.

As the market is advancing your suggested price is rather on the low side. Up to now, we can not **entertain** your **counteroffer**. With reference to the goods you ordered, we have decided to accept your order at the same price as that of last year.

As our stocks are **running short**, we would advise your order as soon as possible.

We trust you will now **attend** to this matter without further delay.

翻譯範例

親愛的大衛：

感謝您八月十日對我們產品的議價。

由於市場看好，貴公司建議的價格偏低。我們至今無法接受你們的議價。關於貴公司訂購商品，我們決定按去年價格接受貴公司的訂單。

因為我們的現貨逐漸短缺，我方建議您盡速下單。

我方相信貴公司會毫不延誤地關心此事件。

關鍵單字

entertain		採納(建議、要求)等、考慮
counteroffer		反提議、反提案、反建議、還價、還盤
例	running short	逐漸短缺
例	running low	逐漸減少
例	running out	逐漸售完
attend		留意、注目、傾聽、專心於、致力於

關鍵片語

Up to now, ~

事到如今，~

形容某一件事情「至今~」時，"up to now" 是很好的敘述方式。"up to now" 是形容事情仍處於「持續」的狀態，表示「從以前到現今一直是如此」的狀態，是帶有負面意思的一種說明。

【例句】

★ Up to now, Jessica is still missing.
至今為止，潔西卡仍舊下落不明。

★ Up to now, we don't know why Mark shot his teacher.
我們至今仍舊不知道麥克為何對他的老師開槍。

★ Up to now, Mr. and Mrs. Jones still live in the city.
瓊斯夫婦至今仍舊住在城市裡。

★ Up to now, I can't forget what he has done to me.
事到如今，我還是無法忘記他曾對我作過的事。

★ Up to now, we still have no explanation for what happened between them.
直到現在,我方對於他們之間發生的事仍沒有解釋。

★ Up to now I still have not read even the first chapter yet.
直到現在,我仍舊連第一章都還沒有讀。

★ Up to now I still have a rotten relationship with my brother.
直到現在,我和我兄弟的關係還是很糟糕。

★ I mean, up to now, I still have questions about your plans.
我的意思是,直到現在,我對貴公司的計畫仍有疑問。

Unit 15 取消報價

We regret we are not in a position to accept the order at the prices.

很抱歉，我方無法按此價格接受訂單。

原文範例

Dear Cathy,

We have received your letter of July 2. As **wages** and prices of **materials** have risen considerably, we regret we are not in a position to **accept** the order at the prices we quoted half a year ago. I am afraid we have to make a new offer for you. In view of the long **continuance** of your **patronage**, we shall do our **utmost** to do at the limit despite the enhanced cost of production. Please find the attached prices list and our best offer for our products. We hope you will **be satisfied with** our quotation.

翻譯範例

親愛的凱西：

我們已經收到您七月二日的信件。

由於工資和原料價格大幅度上漲，很抱歉無法按我方半年前所報價格接受訂單。有鑑於貴公司長期持續的惠顧，雖然生產成本提高，我方仍將盡力以限價成交。請參考附件價格表和我們所提供最優惠的產品價格。

我們希望貴公司能對我們的報價感到滿意！

關鍵單字

wage	工資、工錢
material	原料、材料、物資、衣料
例 material for a dress
女裝衣料
accept	接受邀請[提案等]
continuance	持續
patronage	惠顧、光顧
utmost	最大限度、極度
例 do [try] one's utmost
盡某人的全力
be satisfied with ~	某人滿意~(某事)

關鍵片語

not in a position to ~

在此立場下，無法 ~

當你「身處於某立場無法從事某事」時，就非常適合使用 "not in a position"。

"position" 是表示「立場」、「處境」、「狀態」的意思，若表示「身於某立場」，則為 "in a position"。

因此，而「身處某立場足以作某事」，只要去掉前方的否定詞 "not" 即可，就成了 "in a position to do something"。

【例句】

★ Grace is not in a position to accept his propose.
  葛芮斯無法接受他的求婚。

★ QCA Motorcar is not in a position to recall their cars.
  QCA 自動汽車(公司)無法回收他們的車。

★ We are not in a position to make a reservation.
  我們無法預約。

★ I'm sorry that I'm not in a position to help you.

很遺憾，我幫不了您。

★ Mr. Black is in a position to cancel this appointment.

布萊克先生在此立場下，必須取消這個約會。

★ We regret we are not in a position to accept the order at the prices.

我方很抱歉，無法按此價格接受訂單。

★ We are not in a position to offer guidance on each individual use.

我方沒有立場提供給單一使用者指導。

★ They are not in a position to comment either.

他們也沒有評論的立場。

★ You are not in a position to say anything.

你沒有發表言論的立場。

★ We are not in a position to analyze your samples.

我們沒有立場分析貴公司的樣品。

Unit 16 代理權談判

We recommend ourselves to act as your sole agent for your camera in Taiwan.

我方自我推薦，擔任貴公司相機在台灣的獨家代理商。

原文範例

Dear Sirs,

BCQ is a reliable company with wide and varied **experience** in the line in Taiwan. Having had an experienced **staff** of sales **representatives** and many **excellent show rooms**, we recommend ourselves to act as your **sole agent** for your camera in Taiwan. **Enclosed** is our Agency **Contract** No. 278 in **detail**.

Thank you for your time, and we are looking forward to hearing your reply as soon as possible.

翻譯範例

敬啟者：

BCQ 公司為一家在台灣信譽可靠的公司，在此行業擁有豐富的經驗。

我們擁有經驗豐富的銷售人員和許多一流的陳列室，我方現今自我推薦作為貴公司相機在台灣的獨家代理商。請詳見附件編號 278 代理權合約明細。

感謝您的撥冗，我方期待盡快得到您的回覆。

關鍵單字

experience	經驗 (由經歷所獲得的知識、能力、技能)
staff	(協助主管、負責人、領導人的)全部工作人員
representative	代理商、代銷商
excellent	優秀的、一流的、卓越的
show room	展示室
recommend	建議
ourselves	我們自己
sole agent	獨家的代理

enclosed　　　　　　使封入、附上、裝入

例 enclose a check with a letter
隨函附上支票一張

---

contract　　　　　　契約、合約、合同

例 a verbal [an oral] contract
口頭約定

例 a written contract
書面契約

例 make a contract with ~
與 ~訂合約

---

detail　　　　　　細節、細事、詳細狀況

例 in detail
詳細地、在細部中

關鍵片語

We recommend ourselves to ~
我們自我推薦作為~

　　當你有自信能做好某事時，就是一種很好的自我行銷術，此時你就可以用 "I recommend my-self to do it well" 的表達方式。

　　"recommend" 是推薦、託付的意思，是一種正面積極的用語，建議您在商業書信中不妨多多利用這個單字，以傳達自信的成就。

【例句】

★ I recommended my soul to God.
　我把靈魂託付給上帝。

★ Chris recommended Jeff as a good cook.
　克里斯推薦傑夫是一位好廚師。

★ Professor Lee recommended a good book to me.
　李教授向我推薦一本好書。

★ I recommend you to try this solution.
　我建議您用這種解決方法。

② ⑤ ⑤

① 撰寫 e-mail 英文商用書信的訣竅 ② 常用商業例句 ③ 常用商業名詞 ④ 商業書信範例 ⑤ 商業常用詞彙

Unit 17 財務信譽諮詢

Any information you may give us will be held in absolute confidence.

任何您所提供的資訊絕對會私下地進行。

原文範例

Dear Sir,

OPC Co., who has **recently proposed** to do business with us, has **referred** us to your bank.

We should feel very much obliged if you would inform us **whether** you consider them **reliable** and their **financial** position strong, and whether their business is being carried on in a satisfactory **manner**.

Any information you may give us will be held in **absolute confidence** and will not **involve** you in any **responsibility**.

We **apologize** for the trouble we are giving you. Any **expenses** you may **incur** in this **connection** will be gladly paid upon being **notified**.

翻譯範例

敬啟者：

OPC公司最近要求和我方能有商業往來，對方指示我方向貴銀行查詢。

貴銀行如能提供我方，有關於此公司是否為一值得信賴的公司、對方財務是否健全、對方業務資格是否符合等事項，我方將不勝感激。

任何您所提供的訊息將完全在秘密中進行，無須貴銀行擔負責任。

對於所造成的困擾我方感到很抱歉，任何因此而發生的費用，經正式通報後，我方將會願意支付。

關鍵單字

recently	最近、新近、近來
propose	提議、建議
refer	詢問、查詢(品行、能力等)、參考、參看
whether	不管、無論、是否
reliable	可靠的、靠得住的、確實的、可信賴的
financial	財政(上)的、財務的、金融(上)的

manner	方法、方式、(藝術等的)個人風格、手法、樣式、(文藝上)風格的矯揉造作
absolute	絕對的
confidence	自信、確信、信心

例 in confidence
私下地、秘密地

involve	包含、包括、涉及
responsibility	責任
apologize	辯解、申辯、道歉、認錯、謝罪
expense	(金錢、時間等的)花費、支出、消費、費用、開支

例 Keep expenses as low as possible.
盡可能減少費用開支。

incur	招致、背負、引起（不悅、抗議、損失、花費等）
connection	聯繫、相關
notify	通知、通報、報告、公告

關鍵片語

Any ~ you may give us will ~
任何您所提供給我方的資訊將~

這是一個非常實用的片語，表示未來的情境，所以是用未來式表示的句型，意旨「您所做的某事，將會造成某種後果」，不管是提供具體的物品、提供無形的幫助所引起的結果等，都適合此種解釋句型。

【例句】

★ Any information you may give us will help us to solve this problem.
  任何您所提供的訊息，將有助於我們解決這個問題。

★ Any help you may give us will be thankful.
  您所提供給我們的任何幫助，我們將不勝感激。

★ Any expense you may charge us will be paid by our company.
  任何您所要求支付的費用，將由我們公司支付。

Unit 18 拒絕對方提出的合作計畫

We regret to inform you that we are not in a position to enter into business relations with any firms.

我們很遺憾地通知您，我們無法和其他公司洽談商務事宜。

原文範例

Dear Sirs,

Thank you for your letter of May 4 proposing to establish business **relations** between our two **firms**.

Much as we are interested in doing business with you, we regret to **inform** you that we are not in a position to **enter into** business relations with any firms in your country because we have already had an **agency arrangement** with Go Go Trading Co., Ltd. in Taiwan. **According to** our arrangement only through the above firm, can we **export** our products to Taiwan.

Under the **circumstances**, we have to **refrain** from transacting with you until the agency arrangement **expires**. Your letter has been **filed** for future reference. Thank you again for your proposal and your understanding of our position will be appreciated.

翻譯範例

　　敬啟者：

　　感謝您五月四日來函欲和我方建立雙邊商務關係。

　　儘管我方對於和貴公司建立商貿關係非常有興趣，我們卻必須遺憾地通知您，因為我方已和台灣的 GoGo 有限公司有代理協議，所以我方無法和任何在貴國的公司有商務關係。根據協議，我方惟有透過上述公司才得以出口我方產品至台灣。

　　在此情形下，我方無法與您合作，直至協議到期為止。貴公司的信件已被存檔以備將來的合作可能。

　　再次感謝您的提議，您能考慮我方的立場，將使我方感激不盡。

關鍵單字

relation	關係、關聯、交情
firm	(泛指)公司、企業
inform	提供情報、給與知識
enter into	從事、討論(細節)、處理、作出(決定、計畫等)、參與
agency	代理權、代理、代辦、代銷處、經銷處
arrangement	安排、協議
according to ~	根據 ~
export	輸出、出口(商品等)
circumstance	經濟狀況、境遇、狀況、境況
refrain	忍住、抑制、自制
例 refrain from 禁止~(某事)	
expire	期滿、屆期、失效合約、權利等)
file	歸檔、編檔保存

關鍵片語

We have to refrain from ~
我們必須禁止 ~(作某事)

　　當你因為某些原因無法接受、從事某些事時，就可以用 "to refrain from~"，表示「我們無法~(作某事)」。

　　通常在 "refrain from" 之後要加「動名詞」或「名詞」，一般人容易誤寫成 "to refrain from to do ~"，使用時不得不注意。

【例句】

★ John has to refrain from visiting his kids.
約翰被限制去探望他的孩子們。

★ The official have to refrain from telling the press thetruth.
官員們被制止向媒體説出實情。

★ I can not refrain from laughing.
我忍不住大笑起來。

★ He decided to refrain from smoking.
他決定不再抽菸。

★ We should refrain from making comments.
我方應當盡量不發表意見。

★ How can I refrain from thinking about him?
我如何能夠不想到他？

★ I will refrain from making any comment at present.
現在我拒絕做任何評論。

★ I will refrain from any unfair competition with him.
我盡量節制任何對他不公平的批評。

Unit 19　久未接獲訂單

We have not received your order since last October.

自從去年十月起，我們就沒有接獲您的訂單。

原文範例

Dear Chris,

We have not received your order since last October. We need to **find out** what the trouble has been. Was our service not satisfactory to you? Or did we do something wrong?

It is our policy to **render** the best service to **customers**. You have always been considered as one of our **regular** customers, for you have given us **remarkable patronage**. We need you and we do not want to lose your business. We hope to have the pleasure of serving you again soon.

Your kind reply will be much appreciated.

翻譯範例

親愛的克里斯：

自從去年十月我們就沒有接獲您的訂單，我們需要查明發生了什麼問題。是我們的服務您不滿意或是我們做錯了什麼事？

提供服務給我們的客戶是我們的宗旨。因為貴公司出眾的業績，被視為是我們長久以來的老客戶。我們需要也不想失去貴公司的生意。我們希望還有機會為您服務。

您的善意回覆，我方將不勝感激。

關鍵單字

find out	查明、瞭解真相
render	給予、提供

例 What service did he render (to) you?
他幫了你什麼忙？

consider	看作、認為

例 I consider him to be very clever.
我認為他很聰明。

regular	定期的、定時的

例 a regular customer
老主顧

remarkable	驚人的、非凡的、顯著的、異常的、出眾的

E-MAIL

2
6
7

① 撰寫。第二英文商用書信的訣竅 ② 常用商業例句 ③ 常用商業名詞 ④ 商業書信範例 ⑤ 商業常用辭彙

| patronage | 資助、贊助、支持、保護、獎勵、關照、(對商店等的)惠顧、光顧 |

關鍵片語

We have not ~ since ~ (時間)

自從~(時間) 我們就沒有~

　　當你很久沒有做某事時，就可以使用 "have not ~"，在其後的動詞要改為「過去分詞」，表示「已經很久沒有從事某事」的意思。

　　通常在這句話後面可以再加上 "since ~(時間)"，以說明是「自從~時間至今」的意思。

【例句】

★ I have not seen my parents since I moved to Taipei.
　自從我搬去台北後，就沒有再見過我的父母。

★ Jason has not called his girlfriend since he graduated from school.
　傑森自從從學校畢業後，就沒有打電話給他的女友了。

★ We have not got any presents from my sister since she got married.
　自從我姊姊結婚後，我就沒有再收到她的任何禮物了。

Unit 20 訂單催促

As we are booking heavy orders, we would advise your order without loss of time.

因為我方的訂單量非常大，我方建議貴公司毫不猶豫地立即下訂單。

原文範例

Dear Jason,

We thank you for your enquiry of September 25. Enclosed please find our price quotation.

We have to **say** if it had not been for our good **relationship**, we wouldn't **offer** at this price. We are sure that the prices will **meet** your requirement. We are ready to be at your service and **await** your order.

As we are **booking** heavy orders, we would **advise** your order without loss of time.

Please let us know **as soon as possible** if our offer does not **contain** what you want in order to send you **another** new quotation.

翻譯範例

親愛的傑森：

感謝貴公司九月廿五日的詢價。附件為我方的報價明細。

我們必須說明的是，要不是為了我們的友好關係，我們是不願意以這個價格報價。我們確信此價格符合您的需求。我們已準備好為您服務也正等待您的訂單。

因為我方訂單非常大，我方建議貴公司毫不猶豫地立即下訂單。

如果我方提供的價格不符合您所期望的，煩請盡速告知，以便提供給您另一個新的報價。

關鍵單字

say	說明、建議
relationship	關係、關聯
offer	提供、供給
meet	遇見、符合
await	等待(批准等)
book	預定、預僱、預先安排
advise	建議

② 撰寫 e-mail 英文商用書信的訣竅

② 常用商業例句

③ 常用商業名詞

④ 商業書信範例

⑤ 商業常用詞彙

as soon as possible
盡快地、越快越好

contain　　　　　含有、包含、等於、折合

例 A pound contains 16 ounces.
1磅為16盎司。

another　　　　　又一、再一、另一(個、人、
　　　　　　　　　事物等)

例 in another six weeks
再過六個星期

關鍵片語

As ~, I would advise ~
因為~，我建議~

　　當事情因為某些先天的、特殊的原因導致目前的現況時，就非常適合用 "as ~, ~" 的句型，"as ~" 後面是說明原因，第二個句子則為結果，是你所建議的事實陳述。

　　這個句型非常適合用在「說服」的情境中，所以若是你期望對方盡快下訂單時，就可以用這個 "as ~, we would advise ~" 的句型，會讓對方有一種你是站在他的立場為他考慮的貼心感。

【例句】

★ As it's getting late, I advise you to go.
因為越來越晚了，我建議你去。

★ As they don't care about you, I advise your leaving.
因為他們不關心你，我建議你離開。

★ As Dr. White's promise, I advise you that you should give him a call.
因為懷特博士的承諾，我建議你打電話給他。

Unit 21 確認合約

Please check all the terms listed in the contract.

請您檢查一下合約的所有條款。

原文範例

Jeff,

I'm glad that we've **agreed on** terms of the agency after the past four months.

Enclosed is our Agency Contract No. 278 in **duplicate**, a copy of which please **sign** and **return** to us for our **file**. Please **check** all the terms listed in the contract and see if there is anything not in **conformity** with the terms we agreed on.

Can you **speed up** the contract and let us have it next Friday?

Your special attention to this will be highly appreciated.

翻譯範例

傑夫：

很高興在過去四個月中，我們就代理權一事取得了一致的意見。

茲附上我方代理權編號 278 號合約，一式兩份，請署簽其中一份，並將它寄回我方存檔。請您檢查一下合約的所有條款，看看是否與我們達成協議的條款有不相符的地方。

你們能不能加快簽訂合約的速度，下星期五就給我們呢？

承蒙特別關照此事，我方將不勝感激。

關鍵單字

agree on	同意、應允

例 I agree with you on that question.

在那個問題上我與你意見一致。

duplicate	相同的東西、副本、複製品
sign	簽契約、承擔簽字義務
return	歸、返回、重返

例 return home　回家

file	文件夾、文件箱、訂成冊的文件
check	檢查、確認

①撰寫 e-mail 英文商用書信的訣竅　②常用商業例句　③常用商業名詞　④商業書信範例　⑤商業常用辭彙

| conformity | 相似、符合、適合、一致 |

例 in conformity with ~

和 ~一致

| speed up | 加快速度、加速 |

關鍵片語

Please check all ~

請您檢查一下所有 ~

在商業文書的往返過程中，對於所有的細節問題往往必須要花費相當多的心力在確認的工作上，此確認的工作最常用的句型就是 "please check ~"，所以當你不確定某事時，最好常常利用 "check" 這個字。

此外，除了 "check" 之外，你也可以用 "confirm" (確認) 這個字替代，也是有「檢查」的意思。

【例句】

★ Please check all schedules.
　請檢查所有的行程。

★ Please check all details.
　請檢查所有的細節。

★ Please check all the documents for me.
請為我檢查所有的文件。

★ Check all you can do for me, please.
請檢查所有你可以為我做的事。

★ Would you confirm the flight for me, please?
請幫我確認機票好嗎？

Unit 22 取消訂單

We regret that we have to cancel our order.

很抱歉，我們必須取消訂單。

原文範例

Dear Oliver,

Thank you for your **arrangement**. We have received your goods yesterday. But I'm sorry to say this **quality** is not so **satisfactory**.

I would say that it was unwise of you to supply such **unqualified** goods. We regret that we have to **cancel** our order because of the **inferior** quality of your products.

For all **consequences** arising from **cancellation**, we consider that you are all held liable.

We **recommend** this matter to your prompt **attention**.

翻譯範例

　　親愛的奧利維：

　　謝謝您的費心安排。我們昨天已經收到您的商品。但是很遺憾，這個品質不太令人滿意。

　　我只能說您所提供不符合規格的商品是非常不明智的。很抱歉，由於貴公司產品品質低劣，我們不得不取消訂單。

　　我方認為貴公司對於取消訂單所引起的一切後果，應負全責。

　　我們建議您立即針對此事回覆。

關鍵單字

arrangement	準備、籌備、計畫、預備、安排

例 make arrangements
為~作出安排、與~商定

quality	品質

例 good quality　優質的
例 poor quality　劣質的

satisfactory	令人滿意的、符合要求的、稱心如意的
unwise	不聰明的、不明智的、愚蠢的、輕率的、魯莽的

supply	供給、供應(必需品等)、把~供應給、為~提供
unqualified	不符合品質的
cancel	取消、撤消、廢除、終止
inferior	低級的、較劣的、次等的、品質低劣的、品質較差的
consequence	結果、後果、結局
cancellation	取消、撤消
recommend	建議、勸告
attention	注意、留意、專心
例 pay attention	專注某事

關鍵片語

We regret that we have to cancel ~
很抱歉，我們必須取消~

　　當你因為某原因必須取消某事時，可以使用 "cancel" 這個動詞，這是一個商業及一般口語英文中經常使用的說法。

　　此外，你也可以在此句子之前先說一句 "I regret"，表示你「相當遺憾」的意思，事實上，也許你沒有「遺憾」的本意，但站在禮貌性的人際關係立場上，"regret" 是非常適合表達歉意的用法。

【例句】

★ I regret that I have to cancel our meeting.
很抱歉，我必須取消我們的會議。

★ I regret that Mr. Jones has to cancel this flight.
很抱歉，瓊斯先生必須取消這個班機。

★ She regrets that she has to cancel her wedding.
她很遺憾她必須取消她的婚禮。

Unit 23 船運

I am afraid the middle of October will be too late for us.

十月中旬，恐怕太晚了。

原文範例

Dear Tracy,

Many thanks for your letter of August 15, in which you inform us for the date of shipment.

But I am afraid that shipment by the **middle** of October will be too **late** for us. September is the **season** for this **commodity** in Taiwan market. Would you please **reschedule** shipment to the middle of September **instead**?

We will be obliged if you **exchange** new shipment and look forward to hearing your reply immediately.

翻譯範例

親愛的崔西：

感謝您八月十五日來信通知我方船運的日期。

但是十月中旬交貨，恐怕太晚了。在台灣，九月份是這種商品的上市季節。能請您重新安排船運的日期至九月中旬嗎？

若是貴公司能更改新的船運日期，我方將不勝感激，並期待能盡快聽到您的回覆。

關鍵單字

middle	正中的、中央的、中間的、中級的、中等的、適中的
late	遲的、晚了的、遲於規定時刻的
season	(四季的)季節

例 at all seasons
一年四季、一年到頭

commodity	商品、日用品
reschedule	重新安排時程
instead	代替、頂替、反面

例 Give me this instead.
改拿這個給我。

| exchange | 互換、交換、交易、替代、兌換 |

關鍵片語

I am afraid ~ will be too late for us.

我擔心~(時間) 對我們來說，恐怕太晚了。

　　在人際溝通過程中，「協調」是少不了的一項重要過程。當你參予協調過程時，若有任何你覺得不妥或不適當之處，儘早提出你的問題，將能更有效率提高溝通的品質。如何有效率又兼顧禮儀地提出自己的疑慮？"I am afraid~"是非常好用的句子。

　　此外，若是表示時間或日期的問題，總少不了「太早」或「太晚」的協議，此時 "too late"（太晚）、"too early"（太早）或是 "too rush"（太趕）就是你必須要學會的基本語句。

【例句】

★ I am afraid tomorrow will be too late for me.

我擔心明天對我來說太晚了。

★ I am afraid four o'clock will be too early.
我擔心四點太早了。

★ Perhaps it is too late for us to leave.
或許我們離開太晚了。

★ Don't you think it is too late for the meeting?
你不認為會議時間太晚嗎?

★ It's too late for us to argue now.
我們現在再來爭執為時已晚。

★ I am afraid this date will be too rush for my parents.
我擔心這個日期對我父母而言,恐怕太趕了。

Unit 24 交貨的時間

The earliest shipment we can make is the middle of April.

四月中旬是我方能夠安排的最早的出貨時間。

原文範例

Dear Kathy,

Thanks for all the **assistance** that you have **provided** with **so far**. I'm very sorry, but we really can't **advance** the time of delivery. I'm afraid it won't be ready until the **middle** of April. The **earliest** shipment we can make is the **end** of April. I am sure that the shipment will be made not **later** than the **beginning** of May.

We will be in **touch** with you right after receiving your **thought** to the above questions.

翻譯範例

親愛的凱西：

謝謝您到目前為止提供的各種協助！非常抱歉，但是我方真的無法提前交貨。

恐怕四月上旬之前，無法準備好(出貨)。四月中旬是我方能夠答應的最早的出貨時間。我可以保證，交貨期不會遲於五月上旬。

在得知貴公司對以上問題的看法後，我們將與您聯絡。

關鍵單字

assistance	協助、幫助
provide	提供、供給
so far	目前為止

例 so far so good
目前為止還不錯

advance	使前進、把(工作)推向前進、助成、促進

例 the middle of May
五月中旬

end	在後期的

例 the end of May
五月下旬

earliest	最早的

| later | 較晚的、稍晚的 |
| beginning | 開始、起初、初期 |

例 the beginning of May
五月上旬

| touch | 聯絡、聯繫 |

例 keep in touch with someone
與某人保持聯繫

| thought | 思考、考慮 |

例 take thought
仔細斟酌

關鍵片語

The ~ we can make is ~

我方能夠答應(安排)的~(時間)是~

在商業書信中,當你拒絕接受對方的安排時,能夠同時提出解決問題的方法是最有效率的溝通模式。

因此,在你明確表明時間無法配合時,讓對方在第一時間就知道你所提出的時間建議就非常重要了,因此 " The ~ we can make is ~"(「我方能夠答應(安排)的 ~(時間)是 ~」)就是你必須善用的句型。

【例句】

★ The date we can make is tomorrow.
我們能夠答應的日期是明天。

★ The schedule I can make is not the problem.
我能夠安排的行程不是問題。

★ The wedding she can arrange is the middle of April.
她能夠安排的婚禮日期是四月中旬。

25 提早出貨

When is the earliest date you can ship yourgoods?

你們最早什麼時候可以出貨？

原文範例

Dear Vivian,

I need to discuss the date of shipment with you.

First, when is the earliest date you can ship your goods? **As** the season in Taiwan for apples is in May, could you ship the first consignment by **mid** April?

Second, we find it **impossible** to ask our end-users to **accept** the delayed delivery.

Third, shipment has to be made before April; **otherwise** we are not able to **catch** the season.

Thank you for the trouble you've taken in this matter.

翻譯範例

親愛的薇薇安：

我需要和你討論一下船運出貨的時間。

第一，你們最早什麼時候可以出貨？因為在台灣蘋果的銷售季節在五月，你們能四月中旬送來第一批貨嗎？

第二，我們認為，無法說服我們的客戶接受延誤的出貨。

第三，四月份以前貨必須裝載上船，否則我們就趕不上銷售季節了。

謝謝您為這件事費心了。

關鍵單字

first	第一、首先
as	因為、既然
mid	中間的、中旬的
second	第二、其次
impossible	不可能的
accept	接受、收受
third	第三
otherwise	其他、否則、不然
catch	抓住、趕上(時間)

❹ 商業書信範例

關鍵片語

When is the earliest date you can ~?
你們最早什麼時候可以~?

在商業書信中,「時間」是非常重要的一個關鍵點,當你想要陳述有關時間的問題時, "when is it" 是非常重要的詢問技巧,表示「什麼時候」的意思,若再搭配 "earliest" (「最快」)、 "latest" (「最晚」)的詢問,就能完整表達對時間的掌控。

【例句】

★ When is the earliest date you can finish it?
你們最早什麼時候可以完成這件事?

★ When is the earliest time you can buy it?
你們最早什麼時候可以買?

★ When is the earliest date you should leave?
你應該離開的最快日期是何時?

★ When is the earliest date you can depart?
你可以出境的最快日期是何時?

★ When is the latest time Mr. Kidman can be there?
基曼先生最晚什麼時候可以到達那裡？

★ When is the latest date you can return?
你回來的最後期限是何時？

★ In your opinion when is the latest date you would advise?
以你的意見，你可以提供建議的最後期限是何時？

★ When is the latest date I can apply to get in?
我可以申請的最後期限是何時？

★ When is the latest date I can buy a season ticket?
我可以買季票的最後期限是何時？

Unit 26 商品短缺

We have noticed a discrepancy between our invoice and the quantities you specified.

我們注意到，敝公司發票上的數量與貴公司所提的數量有不符之處。

原文範例

Dear Sirs,

Thank you for your notice No. 1256. We are terribly sorry for the short **delivery** of our **model** A4521.

We have noticed a **discrepancy between** our **invoice** and the **quantities** you **specified**. On the quantities you required, we would **ship** the **replacement at once**. It is **due** to arrive at Hong Kong on September 30, and **confirm** by **return** that goods will be ready in **time**.

Let me **reiterate** our sincere regret regarding these problems. Thank you again for your proposal and your understanding of our position will be appreciated.

翻譯範例

　　敬啟者：

　　感謝您編號 1256 的通知函。我們為短缺的型號 A4521 事宜向貴公司致歉。

　　在調查貴公司所申訴的事件時，我們注意到敝公司發票上的數量與貴公司所需要的數量有不符之處。敝公司會針對貴公司所要求之數量，立即運送替代品。預計於九月卅日抵香港，並以收到確認回覆後，貨物將按時備妥。

　　請允許敝公司重申對此問題的誠摯歉意。謝謝您的建議，對於您體諒我們的立場，我們不勝感激。

關鍵單字

delivery	(貨物、郵件等的)投遞、運送物品、(貨物等的)運送、交貨
model	型號、典型、榜樣
discrepancy	矛盾、不一致、不符合
between	在兩者間、介乎兩者之間
invoice	發票、發單
quantity	量、數量

例 a large quantity of ~
大量的~

例 a small quantity of ~
小量的~

specify	詳細說明、逐一登記、指定
ship	裝運
replacement	置換、代替
at once	即刻、馬上
due	應支付的、到期的

例 be due to
應向 ~付[給]、預計(做 ~)、應結於 ~(的)、由於 ~

confirm	確認、正式承認、再次確定
return	歸、返回、重返
in time	即時
reiterate	反覆做、反覆講

關鍵片語

We have noticed ~ between ~

我們已經注意到~介於兩者之間~

　　"notice" 是「注意到」、「感受到」的意思，不管是具體的察覺或是心靈層面的感受，都可以用 "notice" 表示。而 "between" 是「介于『兩者』之間」的意思，當你想要說明兩者之間的差異時，你必須要使用 "between" 的說明。

　　值得一提的是，若是介於「多數」或「一個群體中」，則要用 "among" 而非 "between"。

【例句】

★ I have noticed the unusual position between us.
我注意到我們之間不尋常的立場。

★ No one noticed the different between Chris and Jeff.
沒有人注意到克里斯和傑夫的不同。

★ My parents have noticed the change between their two different thoughts.
我的父母已經注意到他們兩個不同想法的轉變。

Unit 27 品管不良

Our customers complain that the goods are much inferior in quality to the samples.

我方客戶抱怨貨物的品質遠低於樣品。

原文範例

Dear Jack,

Our customers **complain** that the goods are much **inferior** in quality to the samples. I am afraid it is no easy job for us to **persuade** the **end-user** to buy your goods at this price.

And I believe you shall of course **adopt strict measures** to **control** the quality of your own products. We deeply hope you are fully **committed** to **preventing** this sort of problem.

We should be pleased if you would respond to our **request** at your earliest convenience.

翻譯範例

親愛的傑克：

我方客戶抱怨貨物的品質遠低於樣品。要說服客戶以這種價格購買，對我們來說是不容易的。

而且我認為，貴公司當然要採取嚴格措施來控制你們產品的品質。我們衷心期望貴公司盡全力防止問題的發生。

如果你方能儘早回覆，我方將不勝感激。

關鍵單字

complain	抱怨、向~申訴、控告
inferior	低級的、較劣的、次等的、品質低劣的
persuade	說服、勸說、勸~做某事
例 persuade someone to do something 勸某人做某事	
end-user	終端使用者
adopt	採納、接受
strict	絕對的、完全的、嚴格的、徹底的
measure	程度、適度、範圍、限度、界限

control	控制、操控
commit	委託
prevent	防治、防止、制止
例 prevent from~	制止~
respond	作出反應、回答、作答

關鍵片語

be much inferior in quality to ~

品質較~還低~

　　"inferior" 是「低級的」、「較劣的」、「次等的」、「品質低劣的」的意思，在商業行為中，當商品的品質有瑕疵時，就能使用這個名詞。當你要表現目前所持有的商品和某物作比較時，可以用 " A be much inferior in quality to B" 的片語，A 是瑕疵品的名詞、B 則為比較的標準。

【例句】

★ Your goods are much inferior in quality to my knowledge.
您的商品品質非常低於我的認知。

★ The fruit is much inferior in quality to the sample.
這個水果品質非常低於樣品。

★ The products you supply are much inferior.

您提供的商品品質非常低。

28 回覆不滿意商品品質

Unfortunately, we regret your dissatisfaction with our product.

很不幸地，敝公司很遺憾貴公司不滿意我方的產品。

原文範例

Dear Kenny,

Thank you for your letter of September 30 in which you claim the wrong color of the hat. The **matter** has been given through consideration here.

Unfortunately, we regret your **dissatisfaction** with our product; our **investigation** does not **support** your **claim**.

Consequently, we are not prepared to accept the returned hat and will have no **alternative** but to **insist** on payment of the contracted amount.

We would be happy to **continue** to **cooperate** in any way. And your courtesy will be appreciated, and we earnestly await your reply.

翻譯範例

　　親愛的肯尼：

　　感謝您九月卅日信函所提有關帽子顏色不對的申訴，敝公司已做過全盤的考量。

　　很不幸地，雖然敝公司很遺憾您不滿意我方的產品，可是我方的調查是貴公司的索賠要求無法成立。因此，敝公司並不準備接受貴公司將帽子退回的要求，且貴公司除了必須支付合約上的金額之外，別無他途。

　　敝公司將盡全力與貴公司合作。而對於您的協助我方將感激不盡，敝公司將靜待您的回音。

關鍵單字

matter	事件、(具體的或爭論的)事情、問題
unfortunately	不幸地
dissatisfaction	不滿意、不滿足、不滿的原因
investigation	調查
support	證明、證實
claim	索賠、(根據權利而提出的)要求
consequently	結果、因此、所以
alternative	兩者擇一、可選擇之物、可選擇的方案

| insist | 堅持、強調、硬要、強迫 |

例 I insist on this point.
我強調這一點。

| continue | 持續、繼續、繼續從事(工作、研究等) |
| cooperate | 合作、協力、配合 |

關鍵片語

Unfortunately, we regret ~

很不幸地，敝公司遺憾 ~

"unfortunately" 是一種商業書信中，要說明不幸的、不好的事的用詞，表示「惋惜」的意思，可以讓對方對接下來你要說的事有心理準備。而 "regret" 則表示「遺憾」的意思。通常這個句型是使用在較為「負面」、「遺憾發生某事」的情境中。

【例句】

★ Unfortunately, I regret his death.
　不幸地，我為他的死感到遺憾。

★ Unfortunately, she regrets that she can't help.
　不幸地，她因為幫不上忙而感到遺憾。

★ Unfortunately, I regret to inform you that you are fired.

不幸地，我很遺憾地通知你，你被解雇了。

29 賠償的要求

We would ask you to cover any loss.

我方將要求貴公司負責任何損失的賠償。

原文範例

Dear Jack,

Thanks for the latest catalogue and 20 pairs of running shoes received last Tuesday.

I'm afraid I've got a **complaint** about the quality. The **design**, color and size are not qualified as you **described**. The inferior samples have caused us great **inconvenience**, **necessitating** many **awkward** explanations to our customers.

We would ask you to **cover** any **loss**, which might be **caused** as a **result** of the cancellation of the order.

Thank you for your kind reply on above request.

翻譯範例

親愛的傑克：

謝謝您寄來的最新的目錄和二十雙運動鞋！我們在上週二就已收到這些東西。

恐怕我要抱怨品質問題。樣式、顏色和尺寸都不符合貴公司所描述的。該不良品質的樣品造成的不方便，迫使我方必須向客戶解釋。

我方將要求貴公司賠償由於取消訂單可能招致的損失。

感謝您對上述要求的善意回覆。

關鍵單字

complaint	抱怨、訴苦、叫屈、抱怨的緣由

例 make a complaint
對~提出不滿

design	設計、圖樣、設計圖、模樣
describe	敘述、描寫、形容、說明
inconvenience	不方便的
necessitate	需要、使成為必需、迫使
awkward	棘手的、不便的
cover	(足夠)支付[補償](費用、損失等)、抵償(債務)

④ 商業書信範例

loss	導致賠償要求的事件 (如死亡、傷害、損害等)
cause	成為~的原因、引起
result	結果、結局、後果

關鍵片語

We would ask you to ~

我方將要求~

當你必須對某人提出某項要求時，可以用 "I ask you to do something" 的句型，當然如果要顧及禮儀的要求，就可以加個 "would"，就是 "I would ask ~"，此時的 "would" 是沒有特別的解釋意義。

若是否定句型，則只要在受詞後加 "not" 即可，就成為 " ~ ask you not to do ~" 的句型。

【例句】

★ I would ask you to bring it back to me.

我將要求你將它帶回來給我。

★ We would ask him to make this phone call.

我們將要求他打這通電話。

★ They would ask us to keep the secret.

　他們將要求我們守住秘密。

★ I would ask them not to tell anyone about it.

　我們將要求他們，不要告訴任何人有關這件事。

Unit 30 接受索賠

We are prepared to accept your claim.

我方準備接受您的索賠要求。

原文範例

Dear Scott,

In reply to your letter of October 5, we are terribly sorry to learn of complaints about the damage of the goods and are **prepared** to accept your claim.

According to our **investigation**, the goods have been **heavily saturated** by rain. Please accept our **deepest apology** for any inconvenience this matter has **caused** you. We look forward to the pleasure of working with you again **in the very near future**.

翻譯範例

親愛的史考特：

茲此回覆您十月五日的信函，我方相當遺憾地得知您對於貨物損毀的抱怨，並準備接受您的索賠要求。

根據我方的調查，貨物已遭到雨水嚴重浸溼。請接受敝公司對此事為貴公司所造成之不便的深誠歉意。

我們期待在不久的將來，能再與貴公司合作愉快。

關鍵單字

prepare	準備、預備
according	根據、依據
例 according to ~	根據~事由
investigation	調查、研究
例 upon investigation	
根據調查	
例 under investigation	
正在調查中	
heavily	嚴重地、劇烈地、沉重地
saturate	浸透、滲透
deepest	最深的
apology	歉意、道歉

| cause | 引起、發生 |

in the very near future
在不久的將來

關鍵片語

be prepared to ~

準備要從事 ~

在商業書信中，「隨時讓客戶了解目前的工作進度」是促成成功商業行為的不二法門。

當對方了解你目前所有的工作進度都是按照計畫進行時，自然能夠讓對方感到安心，此時建議您可以使用 " be prepared to ~"（準備要從事 ~）的句型，一方面表現出自己高效率的做事方式，另一方面也可以減少客戶的猜疑心。

【例句】

★ We are prepared to leave Taiwan.
  我們準備要離開台灣。

★ We are prepared to provide quality products at a reasonable price.
  我方準備提供合理價格的高品質商品。

★ She is prepared to cancel their wedding.
  她準備取消他們的婚禮。

★ Dr. Wang is prepared to accept his invitation.
王教授準備要接受他的邀請。

★ Joe said he is prepared to place an order before this Friday.
喬表示,在這個星期五之前,他準備要下訂單。

★ We are prepared to give 20% discount.
我方準備好要提供百分之廿的折扣。

★ We are prepared to deal with each existing problem.
我方準備就每一個存在的問題進行討論。

★ We are prepared to solve these problems immediately.
我方準備立即解決這些問題。

★ I promise that we will be prepared to support your company.
我答應,我方將會準備全力支持貴公司。

Unit 31 立即付款的請求

We request your immediate payment.

我們要求您立即付款。

原文範例

Dear Sirs,

Our **records indicated** that NT$8,000 had not been paid since March. We would say that it was unwise of you to have done that. We **request** your **immediate payment**. Please accept our **deepest** apology for any inconvenience this matter has caused you. We hope that you will make every **effort** to **avoid similar** mistake in our future **transactions**.

We are looking forward to the pleasure of working with you again in the future.

翻譯範例

敬啟者：

我們的記錄指出，新台幣八千元的費用自從三月就尚未結清。我們只能說您的所為是非常不明智的。

我們要求您立即付款。請接受敝公司對此事所造成之不便的深深歉意。我們希望貴公司會盡一切努力去避免未來處理事務時，發生類似的錯誤。

我們期待在不久的將來，能再與貴公司合作愉快。

關鍵單字

record	記錄、登錄、證據、證言、證明
indicate	說、說明、表示
request	要求、請求
immediate	立即的、馬上的、即刻的
payment	付款、支付
deepest	最深的、最誠摯的
effort	努力、奮鬥、勤勉、努力的成果、成就、傑作

例 make every effort
盡一切努力

例 with effort　　　　費力地
例 without effort　　　毫不費力地

avoid　　　　　避、避免、防止~的發生

例 avoid an accident
避免事故

例 I can't avoid saying that.
我不能不說那事。

similar　　　　相像的、類似的、同樣的、同類的

transaction　　(業務的)處理、辦理、處置

關鍵片語

We request ~

我們要求~

當你向對方提出要求時，除了可以用較口語化的 "ask" 之外，另一種商業書信經常用的動詞是 "request"，這裡的 "request" 較 "ask" 來得禮貌而正式些，也比較沒有咄咄逼人的氣勢，而是禮貌性地請求對方遵循某事的意思。

【例句】

★ I request him not to hurt her again.
我要求他不要再傷害她了。

314

★ He requested her to go with him.
他求她一起去。

★ He requested that the error should be corrected.
他要求把錯誤改了。

Unit 32 樣品費催款

However, up to this date, it appears that we haven't received your remittance.

然而，截至目前為止，顯示出我方尚未收到匯款金額。

原文範例

Dear Sirs,

As requested in your letter dated July 6, we sent you the samples via UPS on July 8. When sending you our samples, we also informed you the price with samples.

On July 15 we **submitted** to you a **Debit Note** No.7112-8 for the sample **charges**. **However**, up to this **date**, it **appears** that we haven't received your **remittance**.

I am afraid it's because you didn't receive the debit **at the same time**.

If you have any questions about the amount, please feel free to **contact** us without hesitation.

Thank you for your **cooperation**. Looking forward to your prompt confirmation on **above** request.

翻譯範例

敬啟者：

貴公司七月六日來函索取的樣品已於七月八日以 UPS 寄出。寄樣品的同時，也已一併通知您樣品的費用。

七月十五日我方以帳單編號 7112-8 向貴公司提出要求支付樣品費用。然而截至目前為止，顯示出我方尚未收到匯款額。

我擔心那是因為你未同時收到帳單。假使您有任何有關帳單的問題，請您毫不遲疑地與我方聯絡。

感謝貴公司的合作！我們期待收到貴公司針對以上要求所做出的盡早回覆。

關鍵單字

submit	提交、呈遞、提出(意見等)、委託

例 submit a question to the company
把某一問題提交公司

debit	款項、借項(合計額)
charge	索價、價格、收費

例 free of charge
免費

however	然而、不管到什麼程度
date	日期
appear	明顯、變得清楚、看來好像、似乎
remittance	匯款額

例 make a remittance
匯款、開匯票

at the same time	在此同時、同一時間
contact	與~通訊、與~聯繫、與~往來
without hesitation	毫不遲疑地、無延誤地
above	以上的、上方的

關鍵片語

However, up to this date, it appears that ~

然而截至目前為止，顯示出 ~

　　"however" 是指「不管用什麼方法」，當你向對方表示你已經用盡一切方法努力於某事，而「事實仍舊~」時，就可以用這個字。

　　"appear" 是指「變得清楚」，表示目前的事實已經變得更為清楚、明朗，既能明白表示自己的著急，又能催促對方重視此一問題。

　　"up to this date" （直到如今）也可以用 "up to now" 替代，更能強調「當下」的情境。

【例句】

★ However, up to this date, it appears that she doesn't get his calls.
　　然而截至目前為止，顯示出她尚未接到他的電話。

★ However, up to this date, it appears that they don't visit us.
　　然而截至目前為止顯示出，他們尚未拜訪我們。

★ However, up to now, it appears that only the ladies receive his invitation.

然而，截至目前為止，顯示出只有女士們有收到他的邀請函。

★ However, up to now, it appears that no one can
solve this problem.

然而，截至目前為止，顯示出沒有人可以解決這個難題。

Unit 33 失控的狀況

Things have gone beyond our control.

事情的發展是我方無法控制的。

原文範例

Dear Joe,

We have received your goods last week. But we regret to complain that your **consignment** of goods shipped by m. v. "Sunny" is not of the quality and color of the sample piece.

May we **draw** your immediate attention to this matter? Things have gone **beyond** our control. We hope you would **compensate** us for the **loss**.

Thank you for your cooperation. Your prompt reply will be appreciated.

翻譯範例

親愛的喬：

我上星期已經收到貴公司的商品了。但是你方由「陽光」號輪所裝商品的品質和顏色與樣品不符，我們遺憾地對此表示不滿。

您是否能注意此項事件？事情的發展是我方無法控制的。我方希望貴公司能賠償我方損失。

感謝貴公司的合作。您的盡速回覆，我方將不勝感激。

關鍵單字

consignment	托運
draw	引起、招引、激起
beyond	超出、超越、超過~的範圍

例 beyond one's belief
令人無法置信

例 beyond one's control
超出某人的能力範圍

例 It's beyond me.
這個我就不懂了。

compensate	賠償、補償、抵銷、彌補
loss	損失、損失的物、虧損額

關鍵片語

go beyond control

無法控制的、失控的

當事情「失序」、「失去控制」時，"beyond"是非常好用的表示方法，"beyond"代表「遠在~之後」、「落後於~之後」，而"control"表示「控制」，因此，「失序」、「失去控制」就是"beyond control"。

此外，"control"也可以用"imagination"(想像)、"plan"(計劃)、"mind"(所想的)等字來取代。

【例句】

★ What happened has gone beyond our control.
所發生的事情，超越我們所能控制的。

★ Things have gone beyond my imagination.
事情超越我的想像。

★ His learning has gone far beyond me.
他的學問已遠遠超過我。

④ 商業書信範例

Chapter
5

商業常用詞彙

Unit 01 外匯詞彙

foreign exchange	外匯
foreign currency	外幣
devaluation	貶值
revaluation	升值
rate of exchange	匯率
floating rate	浮動匯率
balance of payments	國際收支
direct quotation	直接報價
soft currency	軟性通貨
indirect quotation	間接報價
inflation	通貨膨脹
buying rate	買入匯率
selling rate	賣出匯率
fixed rate	固定匯率
paper money system	紙幣制度
International Monetary Fund (IMF)	國際貨幣基金

Unit 02 會計詞彙

accounting system	會計系統
account	帳目
audit	審計
balance sheet	資產負債表
bookkeeping	簿記
bad account	呆帳
cost accounting	成本會計
financial accounting	財務會計
financial forecast	財務預測
financial statement	財務報表
financial report	財務報告
income statement	損益表
return of investment	投資回報
return on investment	投資報酬率
cash flow prospects	現金流量預測
cash flow statement	現金流量表
statement of financial position	
	財務狀況表
tax accounting	稅務會計

assets	資產
cost principle	成本原則
creditor	債權人
deflation	通貨緊縮
disclosure	批露
expenses	費用
investing activities	投資活動
liabilities	負債
negative cash flow	負現金流量
positive cash flow	正現金流量
operating activities	經營活動
retained earnings	保留盈餘
revenue	收入
solvency	償付能力
business entity	企業個體
capital stock	股本
sole proprietorship	獨資企業
corporation	公司
owner's equity	所有者權益
partnership	合夥企業
stockholders	股東

Unit 03 價格條件詞彙

price	單價
price list	價目表
amount	金額
international market price	國際市場價格
FOB (free on board)	離岸價
C&F (cost and freight)	成本加運費價
CIF (cost, insurance and freight)	到岸價
freight	運費
wharfage	碼頭費
landing charges	卸貨費
customs duty	關稅
port dues	港口稅
import surcharge	進口附加稅
import variable duties	進口差價稅
stamp duty	印花稅
commission	佣金
return commission	回扣

price including commission	含佣價
net price	淨價
wholesale price	批發價
retail price	零售價
spot price	現貨價格
current price	市價
indicative price	參考價格
customs valuation	海關估價

Unit 04 貿易保險詞彙

All Risks	一切險
F.P.A. (Free from Particular Average)	平安險
W.A. / W.P.A (With Average / With Particular Average)	水漬險
War Risk	戰爭險
F.W.R.D. (Fresh Water Rain Damage)	淡水雨淋險
Risk of Intermixture and Contamination	混雜、玷污險

Risk of Leakage	滲漏險
Risk of Odor	串味險
Risk of Rust	銹蝕險
Risk of Shortage	短量險
T.P.N.D.（Theft, Pilferage and Non-delivery）	
	竊盜未能交貨險
Strikes Risk	罷工險

Unit 05 貿易詞彙

purchase	購買、進貨
stocks	存貨、庫存量
cash sale	現金交易
bulk sale	整批銷售
distribution channels	銷售管道
wholesale	批發
retail	零售
unfair competition	不合理競爭
dumping	商品傾銷
antidumping	反傾銷
freight forwarder	貨運代理

trade consultation	貿易磋商
partial shipment	分批裝運
restraint of trade	貿易管制
favorable balance of trade	貿易順差
unfavorable balance of trade	
	貿易逆差
special preferences	優惠關稅
bonded warehouse	保稅倉庫
transit trade	轉口貿易
tariff barrier	關稅壁壘
tax rebate	出口退稅
export credit	出口信貸
export subsidy	出口津貼
dumping	商品傾銷
exchange dumping	外匯傾銷
special preferences	優惠關稅
import quotas	進口配額制
free trade zone	自由貿易區
value of foreign trade	對外貿易值
value of international trade	國際貿易值
generalized system of preferences-(GSP)	
	普遍優惠制

most-favored nation treatment-(MFNT)
最惠國待遇

Unit 06 海關詞彙

Customs	海關
customs broker	報關行
customs agent	報關經紀人
customs bills of entry	船隻出入境公報
customs entry	海關手續
customs bond	海關保稅證明書
customs duties	關稅
customs clearing charges	報關費
customs debenture	退稅憑單
customs declaration	報關單
customs drawback	海關退稅
customs examination	驗關
customs permit	海關許可證
customs specification	海關說明書
customs tariff	關稅率
customs union	關稅同盟

customs warehouse	海關倉庫
customs clearance	海關放行
customs liquidation	清關

Unit 07 關稅詞彙

customs duty	關稅
differential duties	差別關稅
variable import levies	差價關稅
export duty	出口稅
export rebates	出口退稅
export finance	出口信貸
export restriction	出口限制
specific duty	從量稅
tariff	關稅稅則
tariff concession	關稅減讓
tariff quota	關稅配額
tariff escalation	關稅升級
tariff level	關稅水平

Unit 08 進出口貿易詞彙

commerce / trade / trading	貿易
inland trade / home trade / domestic trade	國內貿易
international trade	國際貿易
foreign trade / external trade	外貿
import / importation	進口
importer	進口商
export / exportation	出口
exporter	出口商
import license	進口許口證
export license	出口許口證
commercial transaction	交易
inquiry	詢盤
delivery	交貨
order	訂貨
Bill of Lading	提單
marine bills of lading	海運提單
shipping order	托運單
blank endorsed	空白背書

endorsed	背書
cargo receipt	承運貨物收據
condemned goods	有問題的貨物
catalogue	商品目錄
customs liquidation	清關
customs clearance	結關

Unit 09 貿易夥伴詞彙

trade partner	貿易夥伴
buyer	買方
seller	賣方
manufacturer	製造商
middleman	中間商
dealer	經銷商
wholesaler	批發商
retailer / tradesman	零售商
merchant	商人
concessionaire	受讓人
consumer	消費者
client / customer	顧客

carrier	承運人
consignee	收貨人
agent	一般代理人
general agent	總代理人
agency agreement	代理協議
exclusive distribution right	獨家經營權
sole agency	獨家代理

Unit 10 交貨條件詞彙

delivery	交貨
steamship	輪船
shipment	裝運
charter	租船
time of delivery	交貨時間
time of shipment	裝運期限
voyage charter	定程租船
time charter	定期租船
consignor	托運人
consignee	收貨人
lighter	駁船

shipping space	艙位
tanker	油輪
cargo receipt	陸運收據
airway bill	空運提單
original B\L	正本提單
optional port	選擇港
optional charges	選港費
shipments within 30 days after receipt of L/C	收到信用狀後30天內裝運
shipment during January	一月份裝船
shipment not later than Jan. 31st.	一月底裝船
shipment during Jan./Feb.	一/二月份裝船
in three monthly shipments	分三個月裝運
in three equal monthly shipments	分三個月，每月平均裝運
immediate shipments	立即裝運
prompt shipments	即期裝運
partial shipment not allowed	允許分批裝船

Unit 11 簽訂合約詞彙

price indication	指示性價格
reply immediately	速覆
reference price	參考價
usual practice	習慣做法
business negotiation	交易磋商
without engagement	不受約束
business discussion	業務洽談
time of validity	有效期限
valid till	有效至
purchase contract	購貨合約
sales contract	銷售合約
purchase confirmation	購貨確認單
sales confirmation	銷售確認單
general terms and conditions	一般交易條件
subject to prior sale	以未售出為準
subject to sellers confirmation	需經賣方確認
subject to our final confirmation	需經我方最後確認

脫口說英語：基礎篇

臨時需要說英語，也能輕鬆脫口說英語!
Chapter 1 生活常用　　Chapter 2 辦公室
Chapter 3 電話　　　　Chapter 4 購物
Chapter 5 人際關係　　Chapter 6 客套短語
Chapter 7 交通　　　　Chapter 8 問路
內文規劃八大情境會話，並編撰「延伸用法」
及「相關用法」兩大單元，幫助您記憶學習。
沒有冗長的使用解釋，只要您跟著學習光碟開

口念，就能應付需要簡單對話的場合!

菜英文-實用會話篇

別以為說英文是出國遊學者的專利，
只要看得懂中文，
你也可以開口說英文!

用中文學英文
中文解釋→ 英文片語 → 中文發音
循序漸進學習英文片語會話!

菜英文-生活基礎篇

超強英文學習手冊!
顛覆語言學習障礙!!
會說國語＝學會英語!

你曾經逃避說英文嗎?
本書保證讓您在最短的時間內，就可以開口
說英文。
在必須說英文的場合中，不要再當個說不出

話的沈默者，哪怕只是說一句I see!都可以化解彼此的尷尬。

永續圖書
線上購物網

www.foreverbooks.com.tw

◆ 加入會員即享活動及會員折扣。

◆ 每月均有優惠活動，期期不同。

◆ 新加入會員三天內訂購書籍不限本數金額，
即贈送精選書籍一本。（依網站標示為主）

專業圖書發行、書局經銷、圖書出版

永續圖書總代理：
五觀藝術出版社、培育文化、棋茵出版社、犬拓文化、讀
品文化、雅典文化、知音人文化、手藝家出版社、瑛申文
化、智學堂文化、語言鳥文化

活動期內，永續圖書將保留變更或終止該活動之權利及最終決定權。

商業實用英文E-mail‧業務篇

雅致風靡　典藏文化

親愛的顧客您好，感謝您購買這本書。即日起，填寫讀者回函卡寄回至本公司，我們每月將抽出一百名回函讀者，寄出精美禮物並享有生日當月購書優惠！想知道更多更即時的消息，歡迎加入"永續圖書粉絲團"您也可以選擇傳真、掃描或用本公司準備的免郵回函寄回，謝謝。

傳真電話：(02) 8647-3660　　　　電子信箱：yungjiuh@ms45.hinet.net

姓名：		性別：　□男　□女	
出生日期：　年　月　日		電話：	
學歷：		職業：	
E-mail：			
地址：□□□			
從何處購買此書：		購買金額：　　　元	
購買本書動機：□封面 □書名 □排版 □內容 □作者 □偶然衝動			
你對本書的意見： 內容：□滿意□尚可□待改進　編輯：□滿意□尚可□待改進 封面：□滿意□尚可□待改進　定價：□滿意□尚可□待改進			
其他建議：			

總經銷：永續圖書有限公司

永續圖書線上購物網
www.foreverbooks.com.tw

您可以使用以下方式將回函寄回。

您的回覆，是我們進步的最大動力，謝謝。

① 使用本公司準備的免郵回函寄回。

② 傳真電話：（02）8647-3660

③ 掃描圖檔寄到電子信箱：

yungjiuh@ms45.hinet.net

沿此線對折後寄回，謝謝。

2 2 1 0 3

 雅典文化事業有限公司　收
新北市汐止區大同路三段194號9樓之1

雅致風靡　典藏文化